Gifts of the Honeybees

Gifts of the Honeybees

Their Connection to Cosmos, Earth, and Humankind

Karsten Massei

*with illustrations
by Franziska von der Geest and the author*

SteinerBooks | 2022

Copyright © 2022 by SteinerBooks

Published by SteinerBooks | Anthroposphic Press, Inc.
PO Box 58
Hudson, New York 12534
www.steinerbooks.org

This work was originally published as *Die Gaben die Bienen*
by Futurum Verlag, Basel, 2014.

Translated from the German by Ines Kinchen

ISBN: 978-1-62148-310-6

All rights reserved. The words and images that constitute this book may not be reprinted or distributed electronically, wholly or in part, without written permission from the publisher.

Printed in the United States of America

Contents

Preface *by Alex Tuchman*	vii
Introduction	ix
I. Spirits of the Animals	1
II. My Path to Honeybees	11
III. Elemental Beings of the Hive	
The Bienerich	19
Guardian of the Apiary	22
Guardian and Bienerich	23
Guardian and Beekeeper	25
IV. The Honeybee Queen	
Marriage Flight	27
Winter Path	30
Sun Being	35
Relationship to the Colony	36
Relationship to the Human Soul	38
V. The Worker Bee	
Being of a Worker Bee	42
Flight of a Worker Bee	43
Visiting Flowers	45
Soul Being of an Individual Bee	47
Relationship to the Human Soul	49
Effects on Human Sleep	52
Plight of Bees	55
VI. The Drone	
Being of the Drones	59
Drone Congregation Areas	61
Importance to the Colony	63
VII. Threefold Bee Colony	65

VIII. THE BEING OF THE GREAT BEE
    The Great Bee and the Earth   73
    Substances of Honeybees   75

IX. THE HIVE
    Archetype of the Beehive   73
    Birth of a Colony: The Swarm   75
    Swarm Cluster   78
    Moving In   81
    Twelve Aspects of a Bee Colony   82
    Essence of Honey   87
    Bee Venom   90
    Honeycomb   92
    Warmth   94
    Spiritual Earth   95

X. TRANSFORMATIVE WORK OF BEES
    Transformation of the Earth   97
    Human Sleep   103
    Realm of the Dead   104
    Animals of the Threshold   105
    The Varroa Mite   107

XI. HUMAN-BEE AND BEE-HUMAN
    Soul of a Beehive   109
    Winter Cluster and the Human Soul   110
    The Honeybee Path of the Soul   112
    Gifts of the Honeybees   116
    Words of the Fairies   117
    Worker Bees' Spiritual Gifts to the Beekeeper   120
    Beekeepers and Their Colonies   122
    An Image of the Future Soul   124

XII. HONEYBEE VERSES   127

XIII. MESSAGES FROM HONEYBEES   143

    About the Illustrations   *159*

    Notes   *163*

Preface

Humanity's situation on Earth today can be intimately felt by those who are still closely tied to nature and carefully observing the phenomena of our natural world. Even ideal environments with plenty of healthy forage do not render pollinators immune to the effect human beings are having on our global ecosystem. There is no longer any place that has not been touched by the results of modern humanity's estranged relationship with our Mother Earth.

In filling the whole Earth with our presence, we also are met with the responsibility of consciously caring for its diversity. In this challenge, the bees can rise before us as wonderful teachers. Resilience and adaptability are certainly two outstanding capacities that the bees embody.

In partnering and living with the bees, we have observed time and again new phenomena that we have never seen or heard of before. The bees manifest their wisdom, instincts, and being in patterns that we can recognize, but they seem to reflect the wisdom of the planetary spheres—the patterns and relationships are constantly in motion and dynamic and never exactly repeat themselves. The bees are ever-new in their creative interaction with the world, and time and again we are amazed at what they are capable of. Creative adaptations, uncommon death processes, cross-colony partnerships, and other "out-of-season" anomalies have been incredible teaching tools in helping the mind to stay open, curious, and playful. This brings the creative and challenging situation of having to regularly question one's own knowledge and experience of who the bees are and what they need.

It is in this way that I see how developing a deeper understanding of the honeybees and cultivating a personal relationship with the hives are the most essential tools that a beekeeper has for accompanying the bees into the future and supporting healing, adaptation, vitality, and resilience. A devotion to the bees as our kin, as our partners in evolution, as our teachers on this healing path—this builds the trust that is necessary for love to bloom and grow between honeybees and human beings. We need each other!

It is in this light that I extend my sincere gratitude and joy to Karsten Massei for aiding us in deepening our understanding and our capacities for relating to the honeybees with ever more intimacy and tenderness. This knowledge and inspiration will surely serve the honeybees, humans, and our beloved Earth, and continue to allow for the right actions to be taken in support of the whole.

—Alex Tuchman
Spikenard Honeybee Sanctuary
Floyd, Virginia,
July 2022

Introduction

This book contains the results of research carried out using the means of suprasensible observation. Its aim is to show what can be revealed about the nature of honeybees by this kind of research, which takes into account the realms of life not able to be perceived by the senses. This research is based on a view of the world and its appearance that assumes every sensory phenomenon is an indication of hidden, suprasensible events. The world of the senses is a world that simultaneously conceals and reveals. The spiritual weaving within all manifestations is hidden, but simultaneously reveals itself, because the spiritual world uses the world of the senses to come into appearance. In fact, the physical and spiritual worlds are not separate. Limiting oneself to the facts that can be perceived by the senses therefore means ignoring a substantial portion of reality.

This book is for all people who care about bees, meaning both beekeepers and non-beekeepers. Short explanations are given for non-beekeepers as needed, so that it won't be necessary to consult a manual on beekeeping. I would also like to mention that this book does not discuss any topics related to beekeeping practices. That will hopefully be the subject of a subsequent publication.

Bees are in great danger. Discussions with beekeepers have made this very clear over and over again. There is a lot of research and speculation about the reasons why bees are dying. The causes are very varied, but there is no doubt that it ultimately leads back to humans. They are the actual cause of bees dying. Technical and economic development that began in the nineteenth century has increasingly led to dealing with the beings of nature disrespectfully and in a way that is destructive to life on Earth. This book is strictly the opposite of such an approach. The spiritual beings that are connected with the life of bees, and whose messages are shared in this book, all unequivocally voice that the plight of bees is actually the plight of human beings. In the torments that humans inflict on the creatures of the Earth, a hardship is expressed that can only be our own. The being of the bees provides perspectives that show how the relationship between humankind and the beings of Earth can be understood. and what the tasks of human beings can be to maintain a responsible relationship with the Earth.

When studying honeybees, one will find that they are creatures whose secrets are not easily elicited. A bee colony is an extremely complex entity. It consists of many thousands of individuals who have the ability to respond wisely to a wide variety of circumstances. To this day, it is not clear how they achieve this. In fact, one could say that a bee colony is imbued with an incomprehensible wisdom that encompasses each individual bee and ensures that the colony remains viable despite ever-changing living conditions.

Anyone observing the life of bees encounters a seemingly inextricable multitude of manifestations of life. In a sense, the study of bees is infinite. The many details to be observed make it difficult to read the basic gestures that would enable one to approach the actual nature of honeybees and bee colonies. Bees elude the observer who tries to make their manifestations of life manageable. Nevertheless, certain regularly repeating phases can be distinguished in the development of a bee colony throughout the year. Each one is a grand image that deeply impresses observers and repeatedly confronts them with a great mystery.

This great mystery of the honeybees has lead humankind throughout the ages to regard them with a deep reverence. Bees have always been considered special, sacred animals. This unspoken knowledge that bees are special beings lies in sharp contrast to the shameful realities of today. What has already come to pass for many animals, that it was impossible for them to continue an existence on Earth, is now also threatening bees. The reason that the plight of bees has received so much public attention is not only due to the concrete physical consequences that the extinction of bees would have on humankind. There is something else behind it based on a deeply rooted feeling in the human soul. The soul senses the meaningful connection between itself and bees. Such forebodings find expression in the fairy tale of "The Queen Bee" by the Brothers Grimm. In the story, Simpleton prevented his two brothers from setting a fire under a wild bees' nest in order to get to the honey by saying: "Leave these animals in peace! I cannot allow you to burn them." Later, the honeybee queen that he saved helps him free a castle that was turned to stone by a spell. Today, we are in the position of Simpleton's two brothers, and it would be wise to heed Simpleton's call.

When we study the nature of bees we recognize the need to approach the phenomena of perception in a meditative manner. By doing so, we notice how impressions that we receive through external perception begin to change. They transform into inner images. Eventually we arrive at attributing no less truth to these soul images than to the insights we gain through the external senses. It is clear that in addition to the "language" of the bees revealed through sensory phenomena, there is also an inner "bee language." The bees don't just want to reach people through our outer senses, they want to speak to us through our inner senses as well.

I have done my best to describe the results of this suprasensory research as clearly and comprehensibly as possible.

This is not an easy task. Time and again, I have discovered that common terminology is inadequately suited to depict suprasensible events. Therefore, I advise you to read slowly and carefully. In many cases, it is only through inner activity that the deeper meaning of certain words and sentences will be revealed. This applies above all to the description of the winter path of the honeybee queen, the question of the archetype of the bee colony, and the indications of the transformative work of bees.

Due to the nature of suprasensory research results, the researcher clearly does not expect them to be blindly accepted. Rather, they can be seen as suggestions for one's own reflections and research. The results of suprasensible research ought to have a practical effect. But, this doesn't mean they must be treated as absolute truths. They can only be proven practically if they are tested by those who want to understand them.

Here, I would like to say something about the methods underlying this discussion of the spiritual nature of bees. Looking at phenomena of the sensory world only reveals a partial reality of life to the observer. A significant part of reality remains hidden from sensory perception. When encountering sensory facts, observers reach a boundary. They experience that the things they perceive become silent at a certain point in the encounter. They feel unable to hold themselves back, to take themselves out of the game, so to speak, in such a way that things and beings can begin to speak. They experience themselves as the greatest obstacle to gaining understanding. Once one has reached and experienced this stage, one now has the opportunity to move forward in a meaningful way. What can be surprising is that what is essential is actually not at all hidden behind appearances. Rather, it reveals itself through what our senses convey to us. By closely observing what is experienced through the senses, one realizes that what is perceived through them becomes a whole field of revelations. One only has to observe things earnestly and patiently. Precise consideration of sensory facts that reveal themselves in plants, animals, and minerals, but also in landscapes and even people, is a first and crucial step towards becoming aware of the manifestations of beings. Colors, shapes, movements, and gestures are letters of a language of the things we observe. The meaning of these letters can be surmised, if one transfers individual observations to one's own soul with an inner movement or gesture. The human soul can read phenomena. It is possible for the soul to have a presentiment and finally recognize the context of meaning, of which individual observations are only partial aspects. Contemplating sensory phenomena with patience and perseverance, and letting them resound in the soul, will in time reveal the existence of spiritual mysteries and lead to an understanding of them.

One can learn to become so deeply involved in observation with one's soul and one's thinking that things begin to speak. However, it is necessary to accompany these perceptions intensively with one's consciousness. The foundation of suprasensory observation described in this book is a path of knowledge that leads to revelations of suprasensible phenomena through the refinement of one's capacities of sense perception. Walking this path, one first acquires the ability to perceive those spiritual beings that we call the beings of the elemental world. They live in direct connection with the sensory realm. They are enchanted in this world.

The further path in the development of suprasensory abilities ultimately leads to the perception of spiritual beings that do not directly reveal themselves in the physical world like the elemental beings. These beings are of a higher nature and are superior to the elementals. In order to perceive them, higher soul capacities are required. They are referred to as beings belonging to the astral realm, while elementals reside in the etheric realm. Entering the realm of astral beings, one becomes aware of the group souls of animals. These reside in the astral world, just as we human beings live with our I in the physical world.

Between 2017 and 2019, I visited the United States three times. During these travels, each lasting many weeks, I experienced the wonderful hospitality of Americans. I met people I feel deeply connected to today, friends that created living memories. I am deeply grateful for the happy hours, days, and weeks. During two of these trips, I was accompanied by my friend, a beekeeper and teacher, Ines Kinchen. I would like to thank her for translating this book from German. I would also like to thank SteinerBooks, for publishing this English translation, and all those who made it possible.

Spirits of the Animals

Honeybees are part of the animal kingdom. It is therefore appropriate to share what suprasensible experience has to offer about the beings of animals. Let us look at the tasks of animals in relation to human beings and their development. In order to understand the subsequent inquiries into the spiritual nature of bees, it is important to acquire an idea of the spiritual connections between animals and humans. The deep significance of the transformative work of honeybees will be discussed in Chapter X. Its meaning will elude the reader who has not yet contemplated the spiritual relationship between the animal world and the human world.

Suprasensible observation of various animal beings shows that there is a deep and essential interconnection between humans and the animal world. These two kingdoms of nature are connected by a spiritual bond. When one discovers this bond, it becomes apparent that humanity is supported by animals. Human life is unthinkable without the spiritual relationship to animals. One could therefore say that humans and animals are related as spiritual siblings.

This siblinghood between humans and animals can only be understood if one has the courage to cultivate a way of thinking that is not closed off to the existence of a spiritual world. When we recognize the possibility that the sense world is not the only world we can perceive, the question of the spiritual connections between animals and humans can be considered in a new way.

From a spiritual point of view, animals are very high beings. The individual animal—the bird singing at first morning light, the shy deer grazing on the far edge of the forest, the leisurely gliding snail, the colorful butterfly—are sensory images of much higher spiritual beings. These higher beings create their sensory expressions in the animals that live on Earth and they guard the spiritual archetype of their corresponding animal species. They are beings from the hierarchy of angels, standing far above the individual animal—the bear or the horse, for example—and evade sense perception. Their essence lives through the individual animal. From them emanates a spiritual life stream through which an animal species can come into appearance on Earth. The connection of an animal species to its higher spiritual being can be compared to the forces that individual human beings receive from their I. A higher animal being works into the animals in a manner comparable to how the human I reveals itself in a person. But these higher spirit beings of animals stand above the individual human being. They are much wiser. Individual animals are the immediate children of these angels. And these angels reveal themselves through them, their children, the animals.

Animal beings are also of great importance to human evolution on Earth. As already mentioned, suprasensible observation shows that animals significantly support human development. Every human being is spiritually surrounded by the beings of animals. This is because the human being has a physical body. In fact, the whole collective of higher animal beings has a deep connection to the human body, which means that their influence exists most of all when the soul is incarnated in the body, specifically during its waking state. During sleep the animal beings step back, while angels of plants unfold their effects. The higher animal beings are more effective while humans are awake and the higher plant beings are more effective while humans are asleep. To those who know how to perceive them, the animal beings are therefore much closer to humanity than the beings of individual plant species. It is easier to receive suprasensible impressions of the animal beings than of the plant beings. Animal beings are only separated from the human soul as if by a small step—a small, almost insignificant space. They are very close to us. It seems as if they were ordered to be the direct guardians of human souls.

Animals have an essential task regarding human beings. Through their existence, they relieve us of the work on ourselves and the Earth that we cannot yet muster out of our own forces. They are our representatives in matters that are still too much for us. Animals take on what humans are not yet able to cope with on their own. They shape and maintain soul spaces, which human beings cannot yet create because we have not yet developed the forces to do so. Through their organization, the animals possess these forces and are completely absorbed in their processes. One could say that in animals, in an organic form, we humans can observe our entire being, which we have not yet grasped consciously. In this sense, animals are very clear signs. In animal life forms, human beings can see the powers and abilities that we will one day be able to have control over consciously.

Animals are accompanying humanity through the evolution of Earth. They are the bearers of important abilities and capacities, which they are transferring to human beings during the course of this development. They accomplish this spiritually, which is why it is easy to overlook the fact that this exchange is taking place at all. Animals are gathered around human beings as their soul siblings.

The being of an animal reveals itself to the attentive observer through its ways of life. Every animal has its own specific gestures and moves in a very special way. Studying these and inwardly understanding them creates an opportunity to grasp the essence of an animal. Through an organ of inner spiritual perception, one becomes the animal that one is observing. One gains access to the spiritual being that speaks and works through the animal. The physical is always an expression of spiritual beings that speak through sense perceptible forms and reveal themselves. The spiritual is always directly present in the physical, otherwise it could not come into appearance at all. In animals, their spirit is revealed in the physical form and in the play of gestures in their movements. A dog moves completely differently than a cat. A picture of its movement can give the viewer a deep impression of its spiritual being. In every case, the expressions of animal life are direct signs that open access to the higher animal being. This is much less hidden than one may think. Animals reveal their essence. This does not apply in the same way to the world of plants. The essence of the plants is much more difficult to grasp for the attentive observer. In order to decipher their pictures of growth, a stronger soul force is required. Animals, on the other hand, make much less of a secret of themselves. They willingly reveal themselves to those who seek their revelations.

Only because it forms a unity with other high animal beings can each animal species serve humankind in a spiritual manner. It only develops its siblinghood with human beings through the connections with all other animals. This also applies to bees. We will now turn to some animals that have a special relationship to bees: the whale, the pig, the horse, the cow, the bird, the butterfly, and the wasp. We will demonstrate that spiritual connections exist between animals and human beings.

The animal being of the *whale* is of a very high nature. The whale swims through the world's oceans with a very specific, virtuous task. This reveals itself to those who know how to approach the meaning of their songs. Their singing is an expression of their mission. This singing permeates the entire Earth, not only the seas, but also other elements. In fact, there is no place on Earth that is not spiritually touched by the songs of the whale. An echo of its song also lives in the air, in light, in warmth, and in stones. If one tries to attune oneself to what lives within these songs, one becomes aware that they have a truly essential significance to the cohesion of the Earth. The songs of the whales hold the Earth together. This may be a strange or alienating idea. However, through patient practice, it is possible to develop an inner sense of hearing the all-penetrating whale songs and to grasp their messages. The soul can be trained to listen to whales anywhere on Earth. Their messages can be experienced clearly. What these consist of will not be explained here. Let us just say that the whale speaks, or rather sings, of the force that shall lead to peace among humankind.

Pigs are animals that are deeply connected to the processes of change and transformation. Through their existence, through their way of life, they fully reveal the principle of transformation. This quality is easiest to grasp through their special gaze. It penetrates to the foundation of life, especially that life that is bound, unfree, frozen, or misguided. A courage is expressed in the gaze of pigs, which does not allow them to look past the darkness, past the unredeemed. Suprasensible perception shows us, astonishingly, that through their higher nature they transform the darkness coming from humans. They not only free themselves from it by taking it into themselves, but also transform it for the circle of life in which they find themselves. This transformation means that they redeem the Earth on which they live and in which they root through. It is redeemed through a spiritual substance, which is formed by the transformation they carry out. There truly is a stream of spiritual substance flowing from pigs into the Earth. This happens when they root around in the Earth. They transform the Earth through their digging. In this way they transfer a kind of healing remedy to the soil, through which the piece of Earth on which they live is nourished and healed. This is an elemental event. Wherever pigs reside, a healing stream nourishes the Earth through them, and the Earth is inhabited by a large number of elemental beings.

A requirement for the pigs' transformational work is that they are well kept and cared for by humans and that they are provided with the proper living conditions they need. Otherwise, they must use up too many forces on themselves. People must relieve them of having to look after themselves. Then they can do their work in peace.

The *horse* is a friend of the human being. It has accompanied humankind since ancient times, through the evolution of different cultures. It has a dignity that radiates onto and is transferred into the human being. This influence of horses on humans always exists, regardless of whether there are horses living nearby or not. Because we have shared such a long path together, the spiritual flow of life between horses and humans, our reciprocal influence, is full of life. Through the being of the horse, we absorb and put into effect a high level of vital life energy. The gifts of horses to humankind concern the ability to stand upright among the conditions of Earth. They envelop the spine, forming an egg-shaped aura around it, more specifically, around each vertebra, from which human beings can experience their uprightness as their own inherent force. It is indeed the horse that gives us the gift of walking upright upon Earth. This is the prerequisite for having the ability to think and for utilizing the forces of understanding. This stream of development emanates from the horses' higher being and leads via uprightness to thinking and understanding. In horses lie the forces of intelligence that reveal themselves in human beings. Horses hold them, but do not use them, so that they can appear in human beings through our own strength.

The *cow* is inclined towards the Earth in such a way that it absorbs the whole Earth, or one could just as easily say that the Earth completely absorbs it. Cows merge with the Earth, with the piece of Earth on which they live, and become a part of it. But the cow does so in such a way that it transforms the substances that it absorbs from this patch of Earth. The cow lets these substances—ingested through food, through breathing, and through its senses—pass through its body in such a way that it transforms them; it ennobles them through its being. By way of the cow's digestion, a renewal of the Earth takes place. That is what makes cows so irreplaceable. This transformative work reveals itself concretely in the processes of substances. The Earth renews itself through the cow. But it can only be renewed because the cow is the bearer of that Earth, which we can call the spiritual, transformed Earth. The cow is an animal of the Earth in the sense that it reveals the sacred, transformed, and transforming Earth through its being. This Earth shows itself in the cow through the cow's digestion, but also expresses itself in the cow's entire being. In the dreaming that surrounds the cow, a glow can be perceived that comes from the spiritual Earth, the foundation of the physically visible Earth. This Earth lives in the unique character of this animal. The spiritual being that reveals itself in cows therefore belongs to highly superior animal beings.

Birds are continually born out of light and air. They are the children of these two elements, indeed continuously so. They die into the light and air and are born out of the light and air so long as they exist and fly and sing. Their behavior, their entire existence is an expression of dying into these two elemental forces and being born out of them. They actually live near death. Death surrounds them, they carry death along with them. They fly toward death and keep escaping from it. This is their fate. Birds are truly animals of the threshold. They threaten to disappear at any moment and actually do disappear, that is, die, but are reborn so quickly that one does not even notice that they had died. One can observe that during every moment of their existence, birds are drawn behind the veil of sense reality and then, in the next moment, emerge again from this veil that lies between the world of sense and the world of spirit. Therefore, they are messenger beings for humankind. And they do actually speak to people. Their language is quiet, imperceptible, but not without effect. Birds are whispering, so to speak, and their song is nothing more than a whisper. Their messages come from the realm that humans have just left behind when we are waking up from sleep. Their songs are golden threads that weave together the invisible and the visible realms, so they don't become separated. They bind the visible Earth to its invisible sister, the spiritual Earth. Birds have earned the right to do this, because they live on the threshold of death. This is why they always have two faces, one spiritual and one sensorial. Their song, their flight, their existence, nourish the Earth and its beings. Through the birds, a weaving of light descends onto the Earth and envelops those who are in danger of being without a sheath and in danger of losing their connection to their own spiritual life thread. The birds spin life threads through which such beings, in danger of falling, can find their way back to their spiritual sources of vitality. What comes from the birds falls to Earth, so that what has fallen can find its way back to the light. Birds are messengers of the living light.

Exalted, sublime spirits envelop the *butterflies*. Butterflies are in their protection. Only when one can be as wide and large as the sheath that the spiritual world lays around the Earth, does one get to where the guardian spirits of butterflies can be experienced. As delicate and playful as butterflies are, as light and weightless, so are the beings who spiritually accompany their lives and their flight. A soul that cannot expand to the limits of the visible world, cannot grasp the being of the butterflies. The angels of butterflies live on the border between the sensorial and spiritual worlds, where a spiritual cloak envelops the Earth. What one perceives sensorially of the butterflies is only the tiny physical center of their whole, encompassing being. A light rises from flying butterflies into the cosmos and is received by the angels. They collect this light within their being. The light rises through the Earth's layers of air, whereby it passes through stages of transformation, until it arrives at the edge of the uppermost layer. Souls of the dead and souls descending onto Earth encounter this light ascending from the butterflies. Some are drawn upward by golden threads, the others downward. The stages of transformation of the rising light of butterflies mark the stages of transformation of the coming and going human souls. Through them, the souls of the dead can reconcile with their fate by ascending into spheres of celestial light, where their soul garment is renewed, no longer consisting of physical matter, but now of woven light. But incarnating souls also descend through these layers and within them approach the Earth and their awaiting destiny. This destiny calls them through the stages of transformation in the light that is nourished by butterflies. Butterflies are animals of the sheath, the sheath of Earth. Their delicacy, their colors, and their way of life indicate that they do not hold their spiritual covering close to them, but that through it they form a unity with the Earth. They are one with the Earth in a way that we can only understand if we are able to form ideas and images of the shape of the Earth's spiritual sheath.

In its spiritual being, the *wasp* is closely connected with what human beings experience in themselves as their understanding-thinking, that intelligence which is oriented to the external facts of life. This means that the wasp is a mirror of the intelligence of human beings, an intelligence that exhausts itself in the material facts of life, oriented solely to the sense world. This intellectual thinking is portrayed in the life of wasps.

This can be seen in the example that, contrary to bees, wasps do not have access to warmth processes. Unlike bees, they are not able to create warmth within themselves. A wasp appears cold, and rightly so. It is surrounded by a superficial, cool, silvery sheen, which is in complete contrast to what emanates from a bee, what can be described as the warm glow of the bee being. This contrast between bees and wasps is also visible in the different materials they use to build their nests. While a wasp creates and utilizes a paper-like substance, a honeybee creates and utilizes wax. The difference between paper and wax expresses the contrast between the cool, silvery being of the wasp and the warm, luminous being of the bee.

However, a wasp's work still carries importance for the bee and its tasks. The wasp, as an animal, takes on the strain created by the predominance of intellectual thinking among humans, which burdens bees. Bees are therefore somewhat relieved of it. The wasp is a being that envelops and protects the bee and thus removes some of the negative consequences of cold, intellectual thinking that people now practice more than ever and carry deep into the life processes of Earth. The wasp enables the bee to do its work more freely and less stressed. This undertaking demonstrates that part of the honey that bees create justifiably belongs to wasps.

My Path to Honeybees

At a particular point in my personal development, I made the surprising discovery that I was able to hear answers to questions I asked. I heard these answers as I normally perceive my own thoughts. I have to admit that, at first, I did not trust what I found. It took time until I was able, through careful examination, to be certain that I could distinguish my own thoughts from the answers I was perceiving. Eventually I started asking questions systematically. I was particularly interested in the world of plants and animals and the beings of the elemental world. I wanted to know what kind of spirits inhabit trees and how they behave throughout the year. I asked questions about the beings of certain plants, butterflies, and also, eventually, bees. The answers gave me indications about those beings of the elemental world who work with bee colonies.

I must admit that the answers I was receiving made me uneasy, because they were often very unusual, downright unbelievable. At first, I was unable to determine where the answers were coming from. I simply heard them and wrote down exactly what I perceived in this process. I still "saw" too little in a clear way. When I asked who was giving me the answers, I learned that they were specific elemental beings. It was only over time that I developed the ability to see the beings who were whispering the answers.

At an early stage the situation was such that I got to know a great many details about elemental beings who work with honeybees, without having actually seen any of these beings. I knew that a colony of bees could only survive because certain elemental beings belong to its spiritual

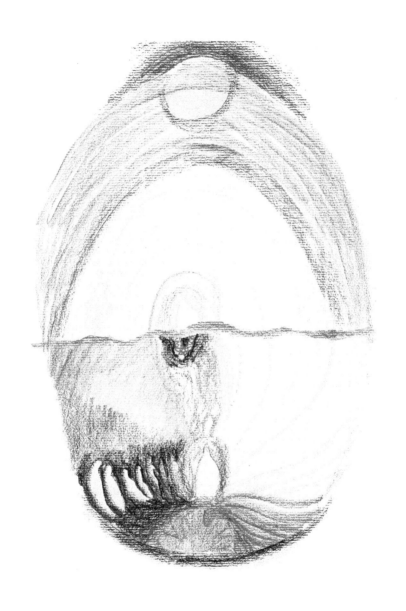

Honeybee Cosmos

organism. I also knew of a being that could be described as the primary elemental being, as it represented all of the elemental happenings in a bee colony. My training, which now began, provided me step by step with the ability to perceive these beings, of whose existence I already knew. However, I remained very skeptical of the answers as long as I could only hear them, and I doubted their accuracy. Even so, I faithfully wrote down what I heard and filled many journals. As I was able to begin seeing elemental beings and come into contact with them, my doubts disappeared because the answers were now confirmed and deepened from another perspective. Looking back, I am amazed at the trust that drove me forward. At first, I did my research without even being fully aware of the scope of what I was engaging in.

Over time I came into more and more contact with beings of the elemental world. I got to know a number of them. The elemental beings of a bee colony were a part of them. This contact helped me see bees from a completely new perspective. I made the acquaintance of the primal elemental being that encompasses all other elementals of a bee colony, and I saw how it resides between the colony and the beekeeper. It ensures that a spiritual space is formed in which human beings and honeybees can encounter each other. This primal elemental being of honeybees eventually received the name "Bienerich."* It was also Bienerichs who

---

* *Bienerich* has etymological roots *Bien* and *erich* (pronounced "bean" and "air-ich," with "ch" as in Scottish *loch*). "'Bien' was a common expression in the German beekeeping literature around 1900, . . . which specifically denotes the honeybee colony as one ('super') organism." Robin F. A. Moritz and Edward E. Southwick, *Bees as Superorganisms: An Evolutionary Reality* (Berlin Heidelberg: Springer-Verlag, 1992), p. 16. *Erich* (from which, English *Eric*) is an Anglo-Scandinavian form of Old Norse *Eiríkr*, from which possibly *Ei* and *rikr*; *Ei*, "one, alone, single" or "eternity, everlasting," and *Ríkr* (later, German *Reich*), "mighty, powerful." Richard Cleasby and Guðbrandr Vigfusson, *An Icelandic-English Dictionary*. 2nd ed. (Oxford: Clarendon, 1957).

explained the cooperation between humankind and beings of the elemental world to me. Bienerichs told me the wishes that the elementals have concerning human beings. You can find more about this in my previous book, *School of the Elemental Beings.*[1] I am always deeply touched by hearing elementals talk about the wishes they have for humankind. What impresses me most is that these wishes are by no means presumptuous or exaggerated. There are many beings of the elemental world who recognize the gravity of the situation in which the Earth finds itself. They suffer from the way humans treat the Earth, and they also know the means by which this plight can be alleviated. It is almost upsetting to hear how easy it would be to implement the methods they suggest for alleviating and remedying this distress. From their point of view, it does not take much of the individual human being to change the state of the Earth, but its necessity must be understood. It comes down to a decision that each person can only make for themselves. The healing of the Earth and social relations can only come from the individual! The elementals I have met agree on this. I have tried to present this point of view in my previous book on the beings of the elemental world. This is their true concern.

There is a meaningful moment that, in retrospect, marks the beginning of my research on honeybees. I had an encounter with a being from the astral realm, who spoke to me in clear words about the plight of bees. This encounter did not take place at a beehive, but in my room at home. It was the first time that I received a visit from a being of this kind. It happened in 2004. I was asked to occupy myself deeply with the being of the bees. Among other things, I heard these words:

> The Earth is in need. She needs your help. She aches under the weight that lies upon her and the burdens that were and still are being imposed on her

by humans. Human beings must create spaces, soul spaces, inner spaces of warmth and light. This is an art, an art of living, the creation of these inner spaces of liberation. Wherever these are not continually created anew, the Earth is breaking. The Earth is expressing her distress, which is actually a distress of the human beings who live on her. The bees work and work to purify the Earth. Because in her there are zones that are stifled, zones of consumption and of desolation. She must be freed from these. Human beings can only assist in this work if they do so from an experience of inner peace, of inner freedom. To assist in illumining the Earth, they must discover their own inner light. Action is necessary.

These words made me take notice. The reason I deeply concerned myself with honeybees for the years that followed has its origins in the content of these few sentences. They formed the seed of my work. I had a wish to understand these statements. I simply knew they were true, but I did not understand them. In the following years I spent a lot of time getting to know the connections that exist between bees, humankind, and the Earth.

My experience was that the phenomena that can be perceived in a bee colony are not easy to interpret. Among the phenomena of life in a beehive, it is difficult to find the gestures through which its being reveals itself directly, as it does in other animal species. I had to learn that a bee colony is an extremely complex being that repeatedly eludes understanding. Patience and perseverance are therefore necessary to understand the being of bees. During my search, three questions came up again and again, which I would like to put forth here before continuing:

- What is the spiritual being of the queen? the worker

bee? the drone? What are their relationships to one another and to the hive?

- How can we understand the wisdom of a beehive? Where does this wisdom come from?
- What significance do bees have for the Earth and humankind?

In order to find answers to these questions, I resolved to let the being of the bees guide me. I wished this being to take me by the hand. Admittedly, the necessary choices that arose were not always easy for me. To let oneself be guided means to take the advice one receives seriously and also to heed it. I have had to learn *a lot* in this regard over and over again. Dealing with beings of the spiritual world confronts us with the task of bringing the knowledge we gain into agreement with our tangible living situation. But this is a path of practice. To this day I feel like a student in this regard—and especially in front of honeybees.

Knowing of the plight the bees find themselves in, I naturally looked for ways to alleviate it. A group of beekeepers from Switzerland, together with the German beekeeper, Margit Fregin, and myself, have made efforts in this direction. But to my surprise I was repeatedly asked by the spiritual beings of the bees to postpone this research and instead do everything possible to deeply understand bees themselves. The rationale was that one can only heal what one deeply understands. I have tried to be faithful to this invitation. That is the reason why there is very little specific information about how to help bees recover in this book. I hope that an understanding of their spiritual nature and the relationships that exist between bees, humans, and the Earth will lead to the creation of curative measures. In recent years I have been able to witness time and again that many beekeepers are already taking steps in this direction.

Through fortunate circumstances, it became possible for me to present the results of my research to more and more interested beekeepers in 2011 and 2012. Each seminar brought me in contact with people who have made it their task to care for and support bees. These include not only beekeepers, but also other people who are interested in bees. From my point of view, along with the sharing of specific research results, it is important that methods are demonstrated and practiced with others. This enables individual people to have experiences that bring them closer to the mysterious bee being. The exercises I suggest serve to internally connect us with specific aspects of the bee being and to gain new insights through this connection. In my seminars, the focus is on meditative work. At first, I was very reluctant to report on my own research because it seemed much more important to me to support the participants in coming to their own insight and spiritual connections of life. I was hesitant to say any more than what the participants could understand through their own experiences. Since then, I have become more generous and forthcoming in this regard. Through the vivid descriptions by beekeepers, the numerous seminars have brought me very close to the world of bees again and again. Each time I have learned many new things. I am very grateful for these encounters and conversations.

The following quote from Rudolf Steiner's lectures about bees given to the workers of the First Goetheanum has accompanied me over the years; it has encouraged me to continue my research, especially during the moments when I thought I was not up for the task:

Insects teach us directly about the highest in Nature.[2]

Spiritual Beings of Bees

Elemental Beings of the Hive

The Bienerich

Time and again, my own path of exploration into the mysteries of the sensory world has led me to encounter beings of the elemental realm. It was also elemental beings of honeybees who first introduced me to the secrets of bees.

Elemental beings maintain and ensoul the life sphere of Earth. Imperceptible to the senses, they inhabit the elements of earth, water, air, warmth, and light. All living creatures, including humans, are dependent on the effectiveness of these beings. One of their tasks is to permeate animals, plants, mineral substances, and human beings with vital life forces. For the bees, this task is fulfilled by the Bienerich.

Every beehive is guarded by an exalted and advanced being of the elemental world. Its presence in a bee colony is necessary for its existence. The presence of that being, the Bienerich, creates a bridge between the colony and the world of essential and invisibly effective life forces. As a being of the elemental world, it brings these forces together, which then become available to the colony. The Bienerich receives the forces necessary for the life of a hive and passes them on to the colony.

In order that the Bienerich can fulfill its task, it offers a large number of very specific elemental beings—necessary for the life of honeybees—the opportunity to live in community with the hive. Therefore, one can imagine a Bienerich as consisting of a great many elemental beings. Through suprasensible observation, one can perceive elemental beings in the hive, which are connected to the colony through the

Bienerich, actually filling it up. A Bienerich makes itself noticeable above all through a certain state of consciousness, by means of which it creates a sheath that the other elemental beings of the hive can be held securely within.

This state of consciousness is the result of a development that preceded its existence as a Bienerich. Through this consciousness, the Bienerich is able to enter into relationship with human beings, especially with beekeepers. This is its second essential task. On one hand, it is so far developed that it is able to form a space that certain elemental beings, absolutely necessary for the life of a hive, can occupy permanently. On the other hand, through its being, it allows the beekeeper and the colony to meet one another, not only on the physical plane, but also in the spiritual realm. The Bienerich is the entity through which the soul of a beekeeper encounters the soul of a hive. This exchange occurs even if the beekeeper is not aware of it. It always takes place when a beekeeper visits their beehives.

For a few years, I worked in intensive collaboration with Annamarie Graf Müller and Beat Müller. At that time, the two lived on a farm south of the area where the three rivers Aare, Reuss, and Limmat flow together, which is also known as Switzerland's "water castle." The farm overlooks the Limmat valley and offers a wide view to the north. These days, the two of them are rarely guests in Switzerland because they live and work primarily in Peru.

Beat had a bee house with about six colonies and also about as many freestanding beehives in his apiary. An incident occurred in this apiary that has special significance to me. Beat, Annamarie, and I were sitting and talking on a bench in front of the apiary on a beautiful summer day. Then I noticed a Bienerich who came over from the apiary and joined us. It sat down on the bench between us and immediately began chatting. I told Annamarie what I was hearing. This was the first time I directly translated what I was receiving from an elemental being for somebody who was present. Later, it happened again and again, but this was the first time. The Bienerich asked the farmer a series of questions that concerned it. It mainly asked about honey. It wanted to know who was receiving the honey that she bottled and sold. All it knew was, some days honey leaves the farm, and it was highly interested to find out what happens to it. Annamarie answered the question as best as she could. The Bienerich wanted to know who was consuming the honey. Annamarie looked at me questioningly. I suggested she talk about customers she knew—who they were, where they lived, and just to give a few brief descriptions. I noticed that the Bienerich only really understood what Annamarie was inwardly imagining. So, I challenged her to describe them vividly. She did, and the Bienerich moved very close to her. It wanted to know what happens to the honey in the households of the people who bought it. Again, I challenged Annamarie to give a very descriptive explanation of how honey is eaten and utilized. The more vivid these images were, the greater was the benefit that the Bienerich experienced.

Through this encounter it became clear to me that a Bienerich can do very little with abstract explanations. It reads the soul of the speaker. It is not words, but inner images through which it understands what is meant. Annamarie and I were both impressed by its desire to know what was happening to honey in such detail. It wanted very specific information. We understood that from its own perspective it could only follow the path of honey up to a certain point. But it was very keen to learn more and to follow the path of honey through to its use. The Bienerich's interest went even further. Finally, it asked about the effects of the honey after it had been consumed by humans. It wanted to know what happened to the honey in the human body. But before we could give an answer, it surprised us with one of its own: "From the perspective of the honey, it is like a kind of beehive, just reversed." We didn't understand what it was trying to say. It continued: "When the honey disappears in a person, it becomes again what it once was. It begins to blossom in the human being." We didn't know how to respond. But these words made us think.

Beat wanted to create a habit of talking to the Bienerichs of his hives more often. I encouraged him to do so and said that he could just talk to them like talking with good friends. "But I don't hear what they're trying to tell me," he said. "It doesn't matter," I said, "you'll find ways to understand one another."

Guardian of the Apiary

Everybody can notice the special mood of grace that can be felt when approaching an apiary. Sometimes this mood can be experienced before one even knows that beehives are nearby. At first, I believed that this was due to an emanation radiating outward from the bee colonies. But then I made the discovery that every place where bees live is guarded by an elemental being.

This elemental being is connected to the bee colonies in such a way that it contains the auras of all the colonies of the apiary within its suprasensible garment. This superior being is continually blessing the colonies. But it also blesses the surroundings and even the people who come visit the bees. Its aura envelops the entire apiary. If one approaches honeybees with awareness, one can experience that at a certain distance one passes through an invisible veil. This is the experience of the suprasensible sheath formed by the guardian of this location.

This great being of the elemental world is one of the innumerable beings who guard all of the various places in the landscape. They are also present in the middle of villages and towns or over lakes and rivers. One can often see them living in large, old trees. They also rise over bridges that form the origin of a settlement. They always have a close connection to human beings. They have followed humankind's development on Earth, and especially within the place they are guarding. This is why they possess a great deal of knowledge about human existence and evolution. These master elemental beings (as they could also be called) serve a higher purpose: the spiritual identity of a place. The other elemental beings of that place are embedded in their being, within the numerous spiritual sheaths of these master elementals.

It is through this guardian being that a place becomes a place of honeybees; a circle of protection is formed around the bee colonies. The hives and the elemental beings associated with them, including the Bienerich, find their place in the landscape through the guardian's existence. Through it, they gain access to the beings of the landscape. The guardian is the gateway to the landscape through the sheath that it provides. Spiritual sheaths in the landscape always have a threefold effect of identity, protection, and incorporation.

The existence of this being confirms the great importance of bees in a landscape. Bees are always creatures of the landscape. One cannot emphasize this connection enough. Bees have important tasks within the landscape, and the landscape has a great influence on them. In the great exchange that takes place between the landscape and bees, elemental beings are, of course, included. They carry out essential tasks within this interaction.

Guardian and Bienerich

The being of a Bienerich has a deep connection to both the bee colony and the guardian of the apiary. One can imagine it in such a way that through the sheath of the guardian, a Bienerich can look into the spiritual reality, the spiritual essence, of the landscape. It is able to see the spirits of plants, trees, water, cliffs, and stones. It can also look at the superior guardians of the landscape, whose mission it is to watch over the individual beings of the elemental world. A Bienerich is able to see tracks of elemental beings left behind on their journeys. For there are many elemental beings who love to wander. Entire clans are on the move. And within this panorama created before it by the sheath of the guardian, a Bienerich also sees the beings who have to eke out their lives on the dark side of existence. These are beings of the elemental world that once fulfilled important tasks throughout the Earth, but were then induced, through human influence, to give up these tasks and now live, unredeemed, in a state of imprisonment.

But how does a Bienerich perceive the human world? What does it learn about people through the sheath, this cloak, of the master elemental? It finds out very little about how humans show themselves physically because a Bienerich is a being without physical senses. It looks into the etheric world. Therefore, it sees neither the physical works of human beings nor their external appearance. But it perceives what radiates from the human world into its own world. It is receptive to the soul, the spiritual of human beings. It is important to know that beings of the elemental world require a special inner strength to look into the human kingdom. Because when they look into the world of humans, they encounter a lot that is hard to endure. Lies, deceit, distortions, and pretense are unbearable for them. They do not understand these soul impulses because these are not in their nature. It takes a lot of strength to look at these and it is very draining for them. At this point I must mention that, more often than preferred, one can hear Bienerichs and other elemental beings grumble about humans because we ask too much of them on account of our immoral nature. Such grumbling they learned from humans themselves.

Of course, a human who turns to nature and treats its beings with reverence and awareness evokes a greater peace and feeling of bliss for beings of the elemental world. As a human, one can make the beings of the elemental realm very happy. This is because the Earth is their body. They are connected to the Earth in the same way we are connected to our body. Therefore, they experience all that human beings do on Earth as if it were happening to their body. Beings of the elemental world take on what people do upon and with the Earth. They are the Earth and they have no choice but to take on what comes toward them from humans. But by doing this, they change. In the worst-case scenario, they are forced to turn away from the original tasks assigned to them. Loving attention is a blessing for them, from which they can be nourished for a long time. Because they spiritually experience the deeds of human beings, these deeds can also nourish them.

Guardian and Beekeeper

The master elemental guarding the apiary also maintains a relationship with the person the bees belong to. Connections of ownership have an important meaning to beings of the spiritual world. To the spiritual world, the moment something belongs to someone, that person is responsible for it. Through the guardian, the soul of the owner, the beekeeper, remains in close connection with the bees. A beekeeper meets the guardian at night in sleep. The spiritual reality of owning land or beehives can be seen in the fact that at night, in the sleeping realm, there are encounters between the owner and the spiritual beings connected to the property. The guardian of the apiary works together with the soul of a human being and places particular qualities into them. One could say that the guardian places bee gold into the sleeping being because it transfers bee capacities. These capacities can be named. They are the soul capacities of inner calm, tranquility, patience to wait for the right moment, and balance between letting go and taking action. These bee capacities that the guardian places into the beekeeper's soul during sleep can be summed up in one sentence: "Half of taking action is letting go and letting go is taking action!" Some beekeepers may find these qualities in themselves. They may have the bees to thank for them.

The Honeybee Queen

Marriage Flight

When one takes on the task of examining the being of a honeybee queen through suprasensible observation, one experiences that she is a very complex being. Her being presents us with many riddles that make it necessary to become fully devoted to her, for she reveals herself only cautiously, and requires a lot of patience from the researcher.

Typically, a honeybee queen leaves her colony once a year by forming a swarm with some of the bees and moving out of the beehive. A new colony emerges from this swarm.

A queen's marriage flight is also of great importance to both her and the colony. During her marriage flight, a queen mates with drones from outside of her hive. Just as with swarming, she leaves the hive while undertaking this once-in-a-lifetime marriage flight. Other than the yearly swarms and onetime marriage flight, she spends her life exclusively within the hive.

The marriage flight takes place in the warm months of the first half of the year, usually around midday. The queen only flies out in sunny weather. The air must be flooded with light. She is accompanied by a small number of worker bees from her own colony and flies to a drone congregation area. On warm days, large numbers of drones from surrounding colonies gather in these areas and await the young queens. When a queen flies by, some drones follow her, attracted by her scent, and mate with her in the air. There are always several drones that mate with the queen. Drones that succeed in doing this die after the process. After mating, the queen

Honeybee Queen Born from the Earth

returns to her colony. She begins to lay eggs within a few days. Her task of laying eggs ceases during the cold season. The colony gathers together to form the so-called winter cluster. As soon as it begins to get warmer again, usually in February, the queen resumes laying eggs.

The marriage flight reveals a honeybee queen's turning toward the light of the Sun. Through this flight she gains the ability to renew her colony. She now becomes the mother of a new colony.

If one pays attention to what is happening suprasensibly during the events of a marriage flight, one can perceive that the queen is not only flying towards the Sun during mating, but rather *into* the Sun. This is actually an astonishing discovery. Spiritually, the queen dissolves into the Sun while mating with the drones. One has the impression that she is being spiritually burned up. Her being surrenders to the light and melts into the sunlight. For some time, the queen's being can no longer be spiritually observed. During this moment, the queen is married to the spiritual archetype of the honeybees. To do this, she must first physically unite with the drones. This union enables the queen to surrender to the Sun. The nature of drones is essential for the queen to receive their archetype. At the same time, however, the nature of drones also prevents the queen from giving herself over completely to the sunlight. The drones save her from a Sun death.

The queen's devotion to sunlight indicates that she succeeds in creating a tremendous opening to the cosmos. She expands in such a way that it seems as if she dissolves into the light of the Sun. In this way, she is able to take in that which allows her to become the mother of her colony. She can now begin the work of laying eggs.

Through the marriage flight, the lifework of a queen becomes a reality. Because she surrenders herself completely to the sunlight, she receives her "crown." It is this permeation by the higher being of the Sun that enables her to fulfill her service to the colony.

Winter Path

In the wake of my 2004 encounter with the being that told me about the plight of the Earth, the bees, and of humankind, I received a series of further messages about certain aspects of the being of honeybees. At this point in time, messages of this kind were no longer unusual for me. I had learned how to handle them. I diligently wrote down what I heard spiritually. Over a period of about two years, I collected these messages concerning the spiritual context within which honeybees live. Some of them, like the following, were about the honeybee queen.

> The honeybee queen is a being born from the highest love of the cosmos. A love similar to that for humankind comes to her from the cosmos. The queen loves her bees like the high beings of the cosmos love humankind. She is an innocent, innocent, innocent being. She is the center of love, the model of love, the resonance of love, the harmony of love.
>
> In September, she gives her farewell. Then she leaves. She travels to the mother of the outer Sun, to where there is peace in the Sun. She goes into the peace to witness this peace, which exists there, but which has not yet arrived on Earth. Among humankind there is still a longing for the peace of the outer Sun, which is the same Sun as the inner Sun of the Earth. Oh peace, oh peace, oh humankind's longing for peace!

This peace lives deep in the human soul, it lives hidden, and has not yet been able to be fulfilled. The honeybee queen travels in search of this peace on her way to the inner Sun of the Earth. She follows the path that would fulfill humankind's longing for peace. If human beings could follow the queen, their longing would be satisfied. She travels this winter path, and spiritually guides those human beings who are willing, so that they too can find the path to the eternal land of peace. If human beings follow their longing for peace, then they are following the queen on her miraculous path into the bright night of the winter Earth, on her glorious path to the peace of Earth.

For the bees remaining behind, the departure of their queen feels like the death of their queen. But the light of the Earth shines toward them through her death. Her destination shines upon the bees remaining in the hive. The bees perceive the glow of the Earth. They experience a reflection of the transformation of their queen. Through the physical body of the queen that has remained, they perceive the queen's winter path as an appearance of light within the hive. The queen does not leave her children, even when she moves into the Earth. It is precisely this withdrawal of the queen that creates the condition for a winter cluster to form.

Observed externally, bees in a winter cluster are grouped around the physical body of their queen. But viewed internally, the bees are grouped around the spiritual events that take place with the queen being inside the Earth. Seen in this way, the center of the winter cluster becomes the center of the

Earth. The core of the Earth is gold. The gold of bees, which becomes visible in their processing of substances, is the Earth gold of the Midnight Sun, the Christmas Sun, the Sun of the night.

I received this message during my research work with Markus Sieber. Although it was already mid-September, we found a small swarm. It had clustered on a fence post at about head height. While we were looking at the swarm, the bees revealed their queen for a short period of time. After the swarm had been captured and was resting in a small box next to us, I received the message shared above, through which I found out about a honeybee queen's winter path for the first time. That winter I began to work to understand these words.

After St. John's Day, in midsummer, a colony begins to prepare for the dark time of the year and to save up provisions. When it gets colder, the bees collect to form a winter cluster. The colony contracts. It now consists only of winter bees, which live up until the time that the first bees hatch again in the new year. Before these young bees are ready, the older winter bees do the work necessary to maintain the colony. Winter bees also have the task of protecting their queen during the cold time of the year. The cluster they form serves to protect not only the queen, but also the entire colony, from the great cold. A honeybee queen survives the winter in the midst of the winter cluster, in the care of her colony.

During suprasensible observation, it is unmistakable that the being of a honeybee queen goes through a transformation in autumn. In autumn and the following winter, she does not form the same unity with her body as during spring and summer. I doublechecked this observation many times until I was finally certain. I could see the queen's being separating from her body. Her body remains behind as a shell and is guarded by winter bees while the inner being of the queen slips away and enters another state.

But, after she has left her body, to where does the being of a queen orient herself? What state does she enter into? I tried to follow the queen's path. In time I realized that her being sinks into the Earth. Her soul being detaches from her body and unites itself with the Earth.

In the latitudes of Central Europe, the detaching of a queen's soul being from her body begins to take place in September. Suprasensibly, it can be observed that for the next six months or so her soul being remains connected to the Earth. Only around Easter does she move back into her body, which has been guarded by her colony in the meantime. The soul of a honeybee queen lives within the Earth during the dark half of the year, just as she lives among her colony during the light half of the year. When she detaches herself from her colony in September, she exchanges her home in the beehive for her home in the inner Earth. In spring, at Easter, she rises from the Earth and inhabits her colony again.

As soon as the queen's soul being has detached itself from her colony, she begins a journey that leads her into ever deeper spiritual layers of the Earth. In doing so, she connects progressively with ever more exalted spiritual beings. The deeper she penetrates into the Earth, the higher-ranking are the beings of the spiritual world who receive her. These beings have the task of enlivening and forming the Earth planet from its innermost center.

During this process it can be observed that the beings of separate honeybee queens merge into one unity. This forming entity eventually unites herself with the center of the Earth. This means that she encounters the Midnight Sun. She unites herself with the Earth's living light. Hereby, she is participating in the primal life force of the Earth. One could say that by descending through the spiritual layers of the Earth, honeybee queens take part in the Christmas events of

Earth. During the thirteen Holy Nights between the years, they look into the deep mystery of Christmas on Earth.

Even though the colony is grouped around the remaining physical body of the queen in the hive's winter cluster, this does not mean that these bees have no experience of what is happening spiritually with the being of their queen. The colony watches the transformation of its queen on her way through the Earth. The bees witness this process through the illumination that continuously emanates from the queen's body. Among other things, this illumination causes the colony not to give up the winter cluster. If the winter cluster were to dissolve, it would lose its innermost light. Seen in this way, the winter bees cluster around the center of the Earth. Winter bees lose the soul of their queen, yet gain the light of the innermost Earth.

In the center of the Earth, honeybee queens encounter a being that can be called the Child of the Earth. This exalted being is in turn guarded by high-ranking angels. As the Child is enveloped by the light of these angels, the queens are enveloped by the light of the Child. During the Christmas season, a large number of spirit beings appear before the Child. Meeting the Earth Child, they are illuminated by the inner light of the Earth. They are guided to the Christmas mystery of the Earth and receive a great blessing. This event takes place during the time of the Holy Nights. The spirit beings of plants and animals, but also elemental beings and angels, find their way to the Child of the Earth. The souls of humankind may also come close to the Child during this holy time. They all approach the Child and stand in its light to be filled with its essence.

From the Child of the Earth, the honeybee queens are gifted a message of blessings for the bee colonies, which they then carry back to them. A queen, who rises again to her colony, passes on what she received from uniting herself with the spiritual Earth in this annual cycle.

It should not go unmentioned that the Earth is not only bright inside. Its interior is also inhabited by beings belonging to darkness. The honeybee queen is for the most part protected from them on her way through the Earth. She observes these beings of darkness, but she does not unite with them.

At Easter time, the soul of the queen returns to her colony. My observation has been confirmed again and again that the return of the queen takes place on Easter night. This is a special moment for a bee colony: the departed queen returns.

The queen shares the experiences of her journey through the Earth with her colony. She does so by bringing the light of the inner Sun of Earth into her colony. The colony is aglow in the light that the queen has gathered. Through the returned queen, the colony is immersed in the innermost Earth's living light and unites with it. Every bee is bathed in this light, enveloped by this light. Like a glaze of life, it wraps around the many bodies of the bees. The unity between the colony and its queen is restored, and their bond is strengthened by the light of the innermost, wintry Earth.

This light forms the unity between a colony and its queen. One could say that through this light the bees experience themselves as one cohesive being, as one hive. This unifying light envelops both the individual bee and the entire colony and is renewed annually through the return of the queen.

Sun Being

Suprasensible observation shows that a honeybee queen has a strong affinity for the essence of light and the Sun. She is the bee-being most closely related to the Sun and its rhythms. The rhythm of day and night and the course of the Sun through the year have a great influence on her. One could say that the queen will always be a child of the Sun. We can observe that she follows the course of the Sun through the year. The spirit being of a honeybee queen moves through the zodiac with the Sun. This journey through the world of stars gifts the queen with the grace and strength needed for the tasks she must fulfill. We cannot really grasp the full being of a queen if we do not also gain a feeling for the cosmic expansion of her being. A queen is not only connected with her hive. Her being has a dimension through which she extends into the world of stars. Wishing for a deeper understanding of her being carries the prerequisite of being able to open ourselves to the cosmic influences affecting the bee colony. The honeybee queen is receptive to the essential qualities of the zodiac.

To the queen, the night is like a little winter and the day like a little summer. This means that for her, every night carries an echo of what she experienced in the Earth in winter. If one tries to comprehend what the night is like for a queen, one experiences that to her the Earth appears illuminated. She meets the Midnight Sun of the Earth. A queen is connected to the Sun at night, which, in our sense experience, disappears behind the horizon. Not only does she see it, but she enters into it, she inhabits the Midnight Sun every night. The bee colony is permeated by this nocturnal light through the queen. She undergoes the changes and wanderings of the Sun. The Sun mystery leads to her being.

Relationship to the Colony

A honeybee queen spends most of her life hidden away inside the hive. She depends on the protection of her colony and an appearance outside of the beehive is a rare exception. But that doesn't mean that the cycle of the year is meaningless to a queen. She is simply connected to the rhythm of the year in a different way than the rest of the hive.

The collective behavior of a beehive is an expression of the seasonal rhythm of the year. Through its development, the hive performs a breathing movement corresponding to this rhythm. The course of the year has a different effect on the queen. She is more spiritually affected by the annual rhythm of the Earth. The connection between a queen and her colony is created specifically by her journey through the seasonal cycles, which she carries out spiritually, while her colony is much more closely involved in the material events and processes. On her journey through the year, a honeybee queen builds the spiritual blueprint of what the colony experiences through the course of seasons. This is how a queen and her colony relate to one another. They complement each other in relation to the cycle of the year, with the queen bringing the spiritual foundation on which the colony can develop its life.

The honeybee queen is deeply connected to the light of the Sun. This becomes clear when one follows her, spiritually speaking, on her marriage flight. To serve her colony, she gives herself to the light. Everything she gains through this intimate oneness with sunlight she passes on to her colony, and her egg laying activity forms the basis of life for the hive. But we have also seen that the queen is deeply connected to the light that shines spiritually inside of the Earth. In deep winter, when the colony is in its winter cluster, she is filled with the Earth's living light.

By looking at the queen being's relationship to light in these two directions, one recognizes that the life of a beehive extends between two poles. Because the queen connects annually with the forces and light of the inner Earth and also once in her lifetime with the forces of the outer Sun, a spiritual space is created in which the being of the honeybees resides and evolves. The effect of bees on the Earth and humankind is rooted in the honeybee queen's giving of herself to both the outer, physical Sun and the inner Sun, the Earth's Sun. In bees, in individual worker bees, drones, and the entire colony, this polarity of forces, held by the being of the queen, plays an essential role.

The marriage flight enables a honeybee queen to physically bring forth a colony. She takes in the spiritual archetype of honeybees. Recognizing the quality that a colony gains through their queen's winter path is much more challenging. One can get an impression of this quality if one observes the events that take place between the queen and the bees that constantly surround her, the so-called court. A queen is fed by her colony. She gets her nourishment from special glands of the worker bees of her court. This particular physical stream of food from the worker bees to their queen corresponds to a spiritual stream from the queen to the worker bees. Through this spiritual stream the being of a queen expands over her entire colony. It is the queen being's life-stream that connects itself to the colony. One can perceive it as a special radiance emanating from the queen. With this streaming light the queen responds to being fed by her colony. An individual bee hereby gains unity with the colony. The colony comes together around this illumination of the queen. This nourishing, peaceful, reconciling light fills and shapes the colony. It has its origin in the queen's spiritual path through the wintry Earth.

In my opinion, humankind's attraction to honeybees has a lot to do with this invisible light of a queen. There are people who feel a type of homecoming when they approach bees. They sense something like a motherland near the bees. They are feeling this because what is emanating from honeybees has its origin in what a queen brings back into her colony from the depths of the Earth. It is precisely this invisible light that touches the human soul.

Relationship to the Human Soul

This inquiry assumes that spiritual relationships exist between the human soul and other beings and things. The human soul is not a self-contained entity. It is exposed not only to sense-based influences, but also to a variety of influences of soul and spirit. Every plant, every tree, and also every animal has a very special influence on the human soul. Therefore, one could say that every plant has a certain "place" in the human soul. For example, there is a place in the human soul that can be called the yarrow place, since this is where yarrow can exert its influence. Then there is also a rose place, an arnica place, the place of the ash tree, etc. This means that the relationship between individual plants and the human soul already exists before a person even discovers the plant and its effect. By recognizing its effect, one recognizes the relationship that already exists between the human soul and the individual plant. Healing properties of plants make this relationship very obvious. The same is true of the relationships between animals and the human soul. One can also speak of animal places in the human soul. Each animal, including the worker bee, the drone, and the queen, has a very specific influence within its respective place in the soul.

These soul places reveal the value of their corresponding plants and animals and also their influence on the human soul. By searching them out, one approaches hidden connections that exist in the natural world. The value that the different beings of the Earth hold for one another begins to emerge.

The influences of various plants and animals on human beings can be experienced consciously. We can encounter them by becoming aware of the expressions of plant and animal beings. This requires a precise and empathetic study of our sense impressions. We can assume that the inner secrets of the life of plants and animals are revealed through what is sense perceptible. By attentively turning toward the expressions of plants and animals that are perceptible to our senses, we allow the observed being and our own soul to enter into a connection. By devoting oneself to what is perceived by the senses, an encounter takes place that creates the foundation for truly "touching" the being of another. Attentive observation is something like a gateway through which the being one wants to get to know and one's own soul can find a common language. It is therefore of immense value to take great care in sensory observation.

This first step is then followed by internal processing of what one has experienced. After working with what is perceived through the senses, one can look at the reactions and responses of the soul.

This can be done by inwardly moving into the spiritual stream that flows from the observed being into the human soul. Hereby, the soul enters the spiritual sphere, the sheath of that being. The soul "wishes" itself, so to speak, into the other being. Here, too, the person must remain observant. Consciously giving oneself over to the influence of the other being, one now perceives the impulses and reactions that arise in one's soul. Practice, care, and perseverance are certainly required to receive reliable results. But it is possible to let go of one's everyday concerns and to allow an inner, spiritual space to open, through which one can come into contact with the spiritual realities of life. This inner deepening creates soul spaces by means of which spiritual beings can communicate in their own way.

If the being of a honeybee queen is approached in this manner, one discovers her soul place, or rather, her soul sphere. One has the experience that she shines over the soul, that she envelops the soul. This is not the case with all beings that people encounter in their soul spaces. The influence of the honeybee queen has a distinct position.

The following verse describes this influence:

A honeybee queen, through her existence and being,
by bees becoming what they are through her,
points to a force
that lives in the human soul.
A honeybee queen, in outer life,
is the appearance of a force,
which enables the soul
to become a unified being.

The Worker Bee

Being of a Worker Bee

The being of a worker bee is deeply connected to the streams of substances that flow through a colony. It is the activity of worker bees that keeps these substance streams alive. The workers collect, carry, transform, and refine the substances that a colony needs. Honey is one product of their activity. But they also create wax—the building material of honeycomb—by sweating it out of their abdomens. Pollination of flowers also occurs through the activity of worker bees. There are truly considerable streams of substances that each colony must work through. A colony gains weight by storing honey, often ten to twenty kilograms (about twenty-two to forty-four pounds). The total amount of honey transformed each year is much higher, up to three hundred kilograms (about six hundred and sixty pounds).

Every gram of honey comes from nectar that is collected from flowers and carried to the beehive. There it is subjected to a process in which the worker bees as a whole are all involved, and through their activity the nectar is transformed into honey. The honey's soothing and healing properties are due, among other things, to this activity of the worker bees.

All bees of a colony, including drones, are descendants of a queen. She is the mother of the whole colony. The bees are therefore siblings to one another. Apart from the queen and drones, a bee colony consists entirely of worker bees. They are responsible for all the complex work necessary for the colony's survival and growth. During the course of her life, about six weeks, a worker bee is able to carry out any tasks that arise. She nurtures maturing brood, cleans the hive, builds new wax comb, ensures the right climate within the hive, accompanies and feeds the queen, attends to the production and care of honey, guards the entrance, and goes on excursions as a forager to gather nectar, pollen, water, and propolis. Until about the twentieth day after hatching, a worker bee is busy with the numerous tasks within the hive. After that, her job is to fly out and bring in the substances the colony needs. It is astonishing how wisely the workers can respond to the demands placed on the colony by the changing conditions in its environment.

The body of a worker bee is capable of producing warmth. That is an amazing fact. The worker bees as a collective can create a specific internal temperature for the hive, which is essential for the development of brood and other activities. This warmth also helps honeybees survive in winter. It is one of the peculiarities of honeybees as there are no other insects that have this ability. The ability to generate warmth is one of the honeybees' most outstanding capacities. Only other higher order animals are able to do this. The fact that a bee colony has its own temperature balance and can autonomously create its warmth tells us a lot about these creatures. It reveals the exalted being of the Great Bee. The fact that worker bees are deeply connected to warmth appears to me as an indication of their true spiritual being.

Flight of a Worker Bee

Observing a beehive, we see right away that the entrance hole is of particular importance. There, the darkness that prevails within the beehive meets the bright outside world. Darkness and light meet. The significance of the entrance hole is emphasized by the observation that the auras of worker bees change when they crawl out and fly upward. The moment they take off, a bright, golden, glowing aura appears. This aura is about the size of a hen's egg and envelops the bees during their flight. Upon their return, as they crawl into the darkness of the hive, this aura disappears again. When a bee is sitting on a flower, the aura also disappears. It can only be observed during the flight of worker bees.

For the worker bee, this aura is a protection that she cannot go without. The aura protects her from the intense impact of light. It may be surprising that the worker bee in particular has to protect herself from sunlight, since she is a being of light. But sunlight would drain the bee's delicate body and cause harm if it were not protected. A worker bee is a being that is particularly sensitive to light. One could say that to her, light is not always the same light. Different qualities of light have different effects on the bee.

Midday light in particular can have a damaging effect on bees' bodies. Evening and morning light are much softer, which is why the aura is more permeable at these times of the day. In weak sunlight, one can hardly perceive an aura around flying worker bees. A resting bee is also not threatened by light. It is only the flying bee that needs protection by this aura.

The sensitivity of a worker bee is surprising. But she is actually an animal that spends only the slightest amount of her life in daylight. Most of the time she is inside the hive among her sisters and close to her queen. It is an exception to be in the light outside of her hive. Seen in this way, the individual bee is actually more of an animal of darkness than of light.

The aura, which can be observed suprasensibly, is created and maintained by the rapid wing movements of a flying bee. The tips of the wings are especially important in this regard. When these are moving, they draw from the formative forces of the light realm surrounding the bee in order to build up her aura. The rapid movements of the wing tips attract the light forces, which flow into the aura and give it a vital and protective form. By moving her wings, the worker bee transforms the light that would otherwise damage her body into a protective covering, making her less sensitive to sunlight. It is precisely through the rapid wing movements that higher spiritual beings contributing to this transformation can become effective.

To understand the being of the bee and her conditions of life, it is therefore advisable to study light and its various qualities. This can help us empathize with the experiences that bees are having within light. Every time of day, every type of weather, and every season has its own light. If one becomes familiar with these different qualities of light, one can learn to see through "bee eyes" and sense what bees are sensing. One will get to know the world as a bee, so to speak.

Among her siblings within a beehive, the individual bee encounters living conditions that are fundamentally different from those that she encounters outside of the hive. In the hive, she is submerged in moving, warm, light-filled activity. Here she is not the single entity she must be when she is flying away from her colony. In the hive, the individual bee is taken in by a life that is familiar to her. The presence of her own queen, the unmistakable scent of her own colony, the security and protection that she can find only here, are experienced by the individual bee when she returns home.

Visiting Flowers

By observing the spiritual relationships of bees, I have been led to the perception of a deep and essential relationship between the flowers visited by worker bees and the beehive, which is the home of the colony and therefore also of individual bees. I observed that a beehive is an inside-out flower and a flower is an inside-out beehive. A beehive is a flower that is turning towards the Earth and growing into darkness. And a blossom is a beehive that opens up to the surrounding light. In a beehive, the blossom dissolves its connection to the light and becomes an inverted chalice, a cave. In a blossom, the beehive is lifted out of the Earth's cohesion and opens up to the realm above ground. Honeybees are drawn to flowers due to the close kinship between flowers and the hive. One can have the impression that flowers and beehives are created from the same spiritual archetypal image, expressing itself in this polarity of forms.

A suprasensible luminance can be perceived within an inhabited beehive. As I have explained, this light is connected to the winter path of the honeybee queen. She is the bearer of this light, which is also the reason why a beehive does not appear dark to worker bees. It is the light of the inner Earth that is perceptible in a beehive, in a colony. It is a light that illuminates the beehive from within. When a worker bee visits a flower, she encounters a being there that is also deeply connected to light. Flowers are beings of light, beings of color. The formative forces of light appear in them. In flowers, the light becomes a living form. In the colorful flowers, worker bees encounter a play of light they are familiar with. It reminds them of the illumination of their queen. They therefore experience flowers as their second home. They have a memory of the beehive they left behind on their flight. So, it is fair to say that bees were created for flowers and flowers were created for bees. This connection only exists due to the kindred relationship between flowers and the hive.

Yet, what urges worker bees to visit blossoms is something else. When a bee visits a flower, she has an experience that does not let her rest, and which induces her to visit the flowers over and over again. With the flowers, she has an experience that she cannot have in the hive among the bees of her colony. By slipping into a blossom, she encounters the

Earth. She encounters the Earth in a delicate manner, but with an intensity that she cannot experience anywhere else in a comparable way, not even among her colony or during her flight. Slipping into a flower and sipping the nectar elicits this experience. Elemental spirits of the plant roots, which connect the plant with the Earth, significantly contribute to this. They mediate a connection between the bee and the Earth. One can observe a special force flowing through the plant when a bee is visiting it and taking in its flower nectar. This awakening stream of life emanates from the roots and root spirits. This creates a kind of sensory experience of the Earth for a bee. An individual bee has a completely different experience of the Earth in the hive among her sisters. There, she experiences the Earth spiritually through the inherent light within. In flowers, on the other hand, she experiences the Earth in a sensory way. This is what makes blossoms so tempting to bees and drives them to visit again and again.

Soul Being of an Individual Bee

When we observe the being of a worker bee with empathetic perception, we discover something astonishing: an individual bee has a wide range of possible experiences that we would not normally expect from an insect, usually only ascribing them to animals of a higher order. It is surprising how an individual bee can differentiate experiences of the events around her. She has an inner life. This possibly strange statement cannot be proven here, but I can simply communicate it as the result of suprasensible perception. Still, I would like to encourage you to imagine yourself in the position of a worker bee and, with an open mind, try to explore the possibilities of her experience. In order to avoid the danger of merely transferring your own sensory experiences to the individual bee, you can repeat the same experiment with other insects. You will discover that butterflies, dragonflies, and flies do not have the same capacity to experience as bees.

An individual worker bee is completely subordinate to the development and needs of her colony. She works entirely for the interests of the colony. Suprasensible observation arrives at the insight that a special soul being lives in every worker bee. One can ask the question of the origin of this being. It turns out that this being had an evolutionary path before it lived in a worker bee. Its origin can be viewed suprasensibly. The soul being of a bee originates from specific elemental beings in the kingdom of plants. It was once the elemental being—more precisely, the flower being—of a plant that grew in untouched, pristine nature.

The flower elemental is chosen to become a bee soul by higher beings instructed by the being of the Great Bee. The flower elemental is essentially "abducted" by these beings from the realm to which it previously belonged. Indeed, it is a process that bears the hallmarks of an abduction. During this process, the flower being loses the characteristics that point to its original ancestry. It is completely transformed. This transformation results in it eventually being breathed into a nascent bee pupa.

During the transformation into a bee soul, the elemental being leaves the spiritual realm to which it originally belonged. It leaves the plant kingdom and enters the soul world of animal beings. Plants are already being touched by the soul realm of animals in their blossoms. In the colors and shapes of their flowers, this soul realm appears in a pictorial, anticipatory form. The chosen plant elemental beings, when being transformed into bee souls, travel a path from plant existence to animal existence. This path of transformation is what we can sense in flowers. Contemplation of this development of bee souls sheds light on the being of honeybees, specifically worker bees. When a worker bee visits flowers on her foraging flights, she comes into contact with beings to whom she is actually related. She encounters a type of being that she herself once was, a being of the elemental world, which she has since left behind to become a bee.

There comes a time when a bee soul leaves the circle of bees again and concludes that existence. Now, she is allowed to go further; she is led back into the spiritual world. Under the guidance of higher beings, she is assigned new tasks. She takes her experiences as a bee with her, just as she took her experiences as an elemental plant being with her into the world of bees. My research has shown that there are several possibilities of further development for a bee soul: she can be given back to certain plants, for which she can serve as a blossom being; she can also connect with the world of birds; or fulfill a task in the aura of a bee colony; or transform herself into a fire being, who is entrusted with very specific tasks in the formation of seeds. Further, one can also observe that she is able to connect with specific beings of ideas.

Relationship to the Human Soul

In my research, I am repeatedly confronted with statements from spirit beings that are puzzling to me. It often takes me a long time to understand these riddles. One of these mysteries is expressed in two compound words that honeybees have continued to say to me: "Bee-human and human-bee." At first, I couldn't really do much with these words. I had a presentiment, rather than an understanding, of their meaning. These two words convey the deep, mysterious relationship that exists between human beings and bees.

In addition to the pair of terms mentioned, another strange statement was spoken to me by Bienerichs, which I did not understand for a long time. It was:

When the bees, the worker bees, fly out, they fly

into darkness. They give themselves to the darkness. To them, the beehive is their home of light. They love it because it illuminates and warms them. In it, they find the peace that they cannot find in the outer light.

What does this mean? We already know that the entrance opening of a hive represents a significant threshold for worker bees. It is truly the case that a bee somewhat loses her cohesion with her colony when she flies out into the bright realm of air. When we see the speed at which bees force their way out into this realm, it becomes obvious that they are crossing an important boundary. On the one hand, it is almost as if the bees are sucked up into the air realm. One has the impression that they await nothing more longingly than to be giving themselves to this space. They take on this task out of devotion to their colony and its survival. On the other hand, one can feel that this is a significant step for a worker bee. It is a kind of threshold crossing.

If one now asks a Bienerich about the nature of this darkness into which an individual honeybee flies, one receives the answer:

> This darkness is yours. It is human made. It is the darkness of your souls that bees must enter when they fly out. To them the air is dark because it is filled with that which streams out as darkness from your souls.

When honeybees break away from their colony, they fly into a soul-filled realm, which rises out from human beings and spreads over the landscape and the various elemental orders. Here we are touching on a fundamental mystery of the being of the bee. Bees directly experience the part of the human soul that lives in the landscape. While flying, they encounter an astrality that characterizes each landscape. This astrality is created, on the one hand, by soul emanations of human beings. On the other hand, the being of every landscape, every location, also has a particular astral quality, just like every flower has a particular color. Bees encounter this astrality directly. One could say that this event, which can be observed from a suprasensible perspective, is an essential key to understanding the being of honeybees. Bees have a strong tendency to strive toward the astrality that they find in the landscape they inhabit. One can imagine these astralities as color beings that lie over every landscape. They can have bright, illumined areas, but also dark, repulsive ones. These latter are filled with lower astrality. Bees experience them as darkness, through which, however, they must fly.

During their excursions, worker bees encounter not only this lower, darker astrality, but also the totality of astral phenomena. It is the collective astral presence that takes hold of the bees when they fly out. Suprasensible observation reveals an event which is difficult to describe. Worker bees, by encountering this astrality, carry out a form of transformation. The lower astrality, which they experience as the beforementioned darkness, leads them into this transformative process. They do not seek out the darkness to lose or exhaust themselves in it, but to transform it. They not only expose themselves to the darkness, but actually contribute to redeeming it. This is the case because an astrality of a higher nature, free from all darkness, is always working through the flying worker bees. The bees not only fly into the darkness, they also fly into the light behind it. Every darkness is always a hidden light. This is the salvation of bees. This is why they do not become paralyzed by their encounters with the forces of unredeemed astrality. They are continually surrounded by the forces through which the transformation process takes place. Every flying bee takes part in the process of transformation through which lower astrality is redeemed.

One could say that a bee's inclination to encounter darkness is due to being permeated by sacred transformative forces. The individual bee, although she flies into the soul darkness of the Earth, remains clearly and firmly in the light of transformation. Her being is permeated with this light. It works through her.

In summary, one could say that bees know the darkness. Their life takes place on the boundary between light and dark. This is why they know the light and the forces of transformation, which are awakened in the moments when light and darkness meet. By encountering the darkness of the soul, honeybees are in danger of dying. But they do not die, because just as they know the darkness, they also know the light, which illuminates and lifts away this darkness. This way, death does not gain power over them. They remain in the life stream of the Earth and can assist in letting the darkness rise up into the light.

Effects on Human Sleep

Here we arrive at an aspect of the bee being that may surprise and create some resistance in the part of the intellect that is used to aligning itself with sensory phenomena. Still, it is important to mention because it demonstrates how bees and human beings are existentially connected. Through indications of this kind, one can begin to guess what is meant by the word pairs "human-bee" and "bee-human."

In most cases, spiritual research begins in a tentative manner. Hints from a world beyond approach the soul, the meanings of which only reveal themselves over time. Understanding often takes years to mature. One must learn to work with preliminary findings and wait for the opportunity to consolidate them. The following information is the result of such an internal process.

In one of my previously published books,* I was able to describe that the foraging flights of bees are a great blessing for the beings of the elemental world. Through these flights, a harmonious weaving of light descends upon the Earth, a kind of resonating dust that is absorbed by the elemental beings of Earth. One could say that they ingest it like a healing substance. This spiritual bee dust is carried down deep into the Earth by elemental beings. It sinks into the Earth. It creates a solid foundation through which the Earth remains open to the future forces of its evolution.

But we can observe that flying bees also create an effect in a different direction. Not only does something fall from them on and into the Earth, but there is also a spiritual effect that rises upward. I have followed this emanation that rises from bees during their gathering flights and have recognized that it penetrates into the regions the human soul traverses during sleep. I observed a living connection between that which arises from flying bees and the existence of the human soul during sleep. This was an astonishing discovery for me. At first, I didn't want to believe that these two things could have anything to do with each other. But, repeated examination showed again and again that this connection does exist.

A sleeping human being has a different relationship to the body than an awake one. In sleep, the soul loosens from the body. It separates and is led to certain realms of the spiritual world. On the way, the influence of the day's events on the soul is increasingly relaxed. The soul becomes more liberated from what affected it while inhabiting the body. More and more, it is withdrawn from the influence of the body and organs. Every night, the soul sheds the tendency to obey the influence of the bodily organs.

Spiritual observation shows that the emanation rising from honeybees affects the sphere of the suprasensible world

---

★ Karsten Massei, *Botschaften der Elementarwesen* [Messages from the Elemental Beings] (Germany: Futurum, 2nd edition, Nov. 2020).

within which the soul has freed itself from the influence of the blood flowing through the human organism. In the waking state, blood circulates through the organism. It is a particularly enlightening experience for an awake person to become aware of the blood that is continuously pulsing through one's own body. It is a tremendous stream of fluid, an unceasing flow. The soul participates in the pulsating and streaming of the blood, albeit unconsciously. It can be observed that by way of a living connection with this flow of blood, the soul opens up to the sphere of the senses, especially through its higher member, the I. Being connected to the blood gives the I of the human being the opportunity to be a part of the sense world. It enables the I to perceive the body it inhabits as well as the external world. Only through the blood does the I gain an experience of its own body and its earthly environment. It perceives both internally and externally.

In a sleeping state, the I must free itself from the binding effect of the sensory sphere. The blood binds the I to Earth, from which it must free itself in order to expand into the spiritual world during sleep. Human beings can only gain experience of the earthly by surrendering themselves to the forces that live in the blood. If one wants to unite with the spiritual world, one must loosen from these forces.

Between falling asleep and waking up, the human soul rushes through the spheres of the planets. It detaches itself from the body and Earth as it travels through the planetary spheres. Each of these spheres indicates a further step in the detachment from earthly influences. Simultaneously, the soul is taken up, step by step, into the spiritual world. Each of the planetary spheres has a specific task that corresponds to each of the individual organs of the body and their relationship to the ascending soul.

The spiritual sphere in which a sleeping soul engages with the forces of the blood is the Sun sphere of the Earth. There the human soul is supported by higher spiritual beings to the extent that it detaches itself from these blood forces. Through one's individual path of development, every human being has a completely unique relationship to the blood forces. One's individuality is reflected in this relationship. We can well imagine that it is not an easy task to break away from these forces. This detachment from the blood forces only happens to a certain degree, which is different for each individual.

Through the flying worker bees a force rises up and reaches the soul of a sleeping human being in the Sun sphere. The bees unite themselves with these souls. They help to dampen the influence of blood forces and support the human I in freeing itself.

What rises into the realms of sleep from the foraging activity of bees during the day helps sleeping human souls achieve a balanced connection to blood forces at night. The blood loses its binding and constricting effect on the soul and I of the human being. This is not to say anything against these binding blood forces, for these, too, are indispensable. Through them, the I can enter into the necessary connection with the physical world. But these forces must not dominate the I. One must be in charge of them, not the other way around. The work of honeybees has an effect of balancing the I and the forces of the blood. This is the healing effect of the flight of bees on humankind. It is not obvious to perceive. It travels in mysterious ways, yet alludes to what is meant by "human-bee" and "bee-human."

Plight of Bees

The living environment of bees has changed greatly due to the technological developments of human beings. One can't help but get the impression that bees are no longer able to cope with these rapid changes, some of which are drastic.

They are overwhelmed by the influences of technology and the many stress-inducing measures taken by humans.

This is not only the case with bees, but also with beings of the elemental world. The life and existence of elemental beings are closely linked to natural processes. To the spirits of plants, animals, water, air, forests, and mountains, the development of the human world has accelerated to such an extent that they are experiencing less and less access to what comes toward them from humans. They were used to living with human beings throughout millennia. This coexistence underwent a radical change in the twentieth century. The sacred threads between human souls and the beings of the elemental world are in danger of being torn.

Let me describe, from a suprasensible perspective, what the burdens on bees consist of today. I am aware that this task must be held in a very serious and responsible manner. These descriptions should not lead to a feeling of powerlessness in the reader. Idleness and resignation would be exactly the wrong things. As a result, bees would lose those humans who actually really want to help them. But to understand bees, it is important to look at their living conditions in a realistic and sober way. For these are also our own living conditions. Everything the bees are showing us affects the human world just as much. I have heard from Bienerichs, time and time again, that if one wants to help honeybees, one cannot ignore the darkness, the forces and influences that are hostile to life. Rather, we are called to look more closely, to perceive these forces and thus be able to assist the bees. Looking away is the worst thing for the bees. They must live with these burdens without having been asked for their consent. A lot is done to bees that they would never choose. Humans become allies to bees when they are not afraid to look into what is dark and hostile to life. Empathy always comes through recognition. It is necessary to have courage to follow the bees to where they come into contact with the hostile influences of modern life.

A worker bee is more exposed to damaging environmental influences affecting bee colonies than drones and queens. On one hand this has to do with the worker bee's tasks for her colony, but, on the other hand, this is also an expression of her being. She is more deeply affected by that which is coming from the periphery because she is naturally more devoted to her surroundings than a drone or queen. The worker bees' tasks cause them to be particularly affected by damaging environmental influences.

The average worker bee in Central Europe exhibits a high level of energetic pollution in her body. Suprasensible observation of the body of a worker bee shows that the wings in particular have lost much of their colorfulness. They should actually be glistening in all colors of the rainbow. A lively, colorful shimmer should emanate from them. In some cases, however, dark, brown-black veils appear over the transparent wings. Energetically, the wing tips are completely shredded and clearly damaged. The fine coat of hair covering her body like fur is also affected. It is often sticky, as if drenched in an oily, dark grease. Her chest and head, eyes and antennae are particularly afflicted. Light emanating from her is faded and dull. If one compares the condition of bees today with that of just fifty years ago, by inwardly going back into the past, it becomes obvious that their luminosity has greatly decreased.

In some places, the strains that bees are exposed to are very high. The situation in Chernobyl is particularly devastating. There I see bees whose aura is very dark. However, it is the bees' task not to leave such places.

What are the forces that lame the aura and body of a bee? Human astrality certainly has a negative impact. But in addition to radioactivity and general pollution and poisoning of the Earth, it is largely electromagnetic radiation in the air that afflicts bees. This change of the environment through human influence puts a strain on their delicate and sensitive bodies. The effects of electromagnetic radiation should not be underestimated. Bees are very exposed to everything that happens in the air. Due to their nature, they give themselves to the air element in a special way. Today, they fly through air that robs them of much strength due to electromagnetic pollution. Instead of nourishing the bees with light and sound, the air eats away at their forces.

It can be observed that these laming, life-sapping forces have an effect all the way into the bees' brood. The pupa, in particular, which already shows clear characteristics of the later bee, is affected by these debilitating forces. They amass around the rear of the pupa. The honeycomb only offers sufficient protection until the pupation phase. The fact that the colony cannot provide the brood with enough protection is, on the one hand, due to the fact that these life-damaging forces are getting stronger. On the other hand, the colony has less and less of its own forces available to protect the brood.

These descriptions naturally raise the question of what options people have to strengthen and protect bees. This is a very important question. Clearly, there are no easy answers. Beekeepers around the world are working diligently and with much concern to find answers. I hope my research may contribute to this pursuit.

The Drone

Being of the Drones

I once heard the following words from a Bienerich: "Once you know the secret of the drones, you will understand honeybees. The secret of honeybees lies in the drones." This is one of the many puzzling pieces of information I have received from Bienerichs.

Male honeybees are called drones. Their job is to mate with the queen during her marriage flight. To do this, drones of various colonies visit special places, so-called drone congregation areas, where they await various young queens. Unlike the worker bees and queen, drones are not tied to a specific bee colony. Although they are born in a colony, they migrate from one hive to another during their lifetime without being rejected. While guard bees rarely let in any foreign worker bees, they let drones pass. This means that drones are a connecting link between the various bee colonies.

Drones do not participate in any of the various tasks necessary for the maintenance of a colony. They don't even contribute anything to their own sustenance. When they are no longer needed to mate with queens, they are driven out of the hive and left to starve. This occurs in late summer. Drones develop in comb cells that are slightly larger than the cells of worker bees. A beekeeper can therefore distinguish between the brood of drones and that of workers. Drones are easy to differentiate from worker bees with the naked eye. Their bodies are bigger and more powerful, and their hum is deeper. And unlike the queen and worker bees, drones do not have a venomous stinger.

The lifeways of drones seem to be polar opposite to that of worker bees. While the workers have a plethora of tasks to attend to, the life of the individual drone appears to be centered around mating with a queen; apparently, this is their only job. Other than that, they lead a rather calm and comfortable life. This is why we can find many descriptions of honeybees in which drones are portrayed as lazy freeloaders.

It is obvious that the being of drones is more difficult to grasp than that of worker bees. In drones one encounters a being that, unlike that of the workers and queen, conceals itself more than reveals itself. They unveil their being much more cautiously than worker bees do. They somewhat deceive the viewer. This cannot be said of worker bees, for they reveal themselves through their activities.

Attention to the nature of drones is an essential prerequisite of being able to understand them. They prove to be defenseless animals with a calm demeanor. The worker bees' strong tendency to keep moving is foreign to the drones. Comparatively, drones can be distinguished by the coherence and unity of their being. They are not driven by the constant urge to work. Their strength comes from the calm and stillness of their being. To immediate observation, the drone can appear as the more mature or fully matured bee.

One gets the impression that a significant part of the drone's being does not appear outwardly. In other words, unlike a worker bee, a drone is in a state similar to human sleep. Their peculiarities are not revealed, but rather remain in a spiritual state, just like in a sleeping person. So, one could say that worker bees embody the daytime side of a beehive's life and drones the nighttime side. During sleep the being of a person is only partially revealed. However, our waking daytime consciousness arises out of this spiritual existence. Without sleep there would be no waking state. Drones are similarly related to worker bees. The sleep of drones is the foundation of the wakefulness and activity of the workers. Only because drones are asleep can worker bees wake and be so effectively active in the colony. Without drones, they would lack peace and calm, which are very important spiritual qualities.

Drone Congregation Areas

Drone congregation areas are special places in the landscape. In his excellent book about bees, Michael Weiler writes:

> For a long time, it was not known why there were drones in a bee colony, and people did not know where they flew to. Reports of so-called "drone congregation areas," supposedly somewhere in the vicinity, were always cropping up without anyone knowing exactly what they were. Most recently it has become possible to ascertain certain facts about these places.
>
> Every year, from the end of April to the end of August, drones from neighboring colonies congregate in a specific area from about eleven o'clock in the morning until around five o'clock in the afternoon. They fly back and forth so fast that they can be hard to spot, and these places are usually only recognized by the low hum of the drones. If a lure is made with a caged queen fixed to a helium balloon attached to a string, the height and extent of such areas can be determined. As soon as she is pulled out of the congregation area, no more drones will follow. In such an artificially contrived way it has been possible to film the mating of a queen.[3]

A queen usually mates with drones from several colonies. Her spermatheca* will then contain sperm from vari-

---

*The *spermatheca,* or sperm pouch, is a sac for sperm storage in the female reproductive tract of various lower animals and especially insects.

ous drones. While queens could also mate outside of a drone congregation area, mating there is much preferred. So, there must be something special about these places.

A drone congregation area is located on a foundation of spiritual forces. This "structure" of forces can be perceived suprasensibly. It creates a space in which a queen can mate. Mating means that the honeybee queen receives the archetypal image of the honeybee. Drone congregation areas seem to offer the ideal conditions for this.

My suprasensible investigation of these mysterious places has not yet been concluded. So far, I have been able to see that a drone congregation area is characterized by the presence of strong Earth forces. A drone congregation area, as a whole, takes on a dome-like shape. One can see that it is permeated by wafting light. High spirits of light are streaming through it. The foundation of forces is continually created from out of the Earth below. It takes its shape from forces of the Earth. Drones fly into this realm of forces and one has the impression that they are swimming in the streaming light. They tumble through the air and light in droves, following the virgin queens who join them. Drone congregation areas exist as long as there are virgin queens. As soon as the mating season ends around July, the congregation areas collapse.

Drone congregation areas are sacred places in the landscape. A special mood can be experienced in these places. One could describe the congregation area as a beehive of drones. It simply does not have a visible form. Congregation areas are dwellings suitable to drones, corresponding to their being. In the physical beehive, drones grow up and are fed. A drone congregation area is the place where they can genuinely reveal their being. The bee colony that raised them is too small for them. They strive for something else. Drones have a more universal character. Unlike worker bees, drones do not belong to a single colony. The unfolding of their being is suited to a spiritual beehive.

Importance to the Colony

As previously described, worker bees have a special relationship with the Sun. This can be seen in the radiance emanating from flying workers, which has an effect in the spiritual realm where the souls of human beings enter during sleep. The effect of this radiance is particularly strong in sunshine. As a result, the soul traversing the sphere of the Sun during sleep is gripped by this effect radiating from worker bees. Again, this has a healing effect.

Drones have no comparable spiritual effect on the human soul. They are all the more spiritually important to the bee colony itself. The special quality the colony experiences through drones is expressed externally in the fact that they are not tied to an individual colony. They form a community of vagabonds traveling between the various colonies. Their "beehive" is very expansive. It encompasses all of the bee colonies in a particular region.

Through their existence, drones transcend the individual colony. Therefore, one can truly look at them as the embodiment of a consciousness extending beyond individual colonies. Through drones, a cosmic quality lives among honeybees, which could not manifest itself through the workers or queen. Drones build a bridge between the colony and the cosmos. It should be noted that drones are not turning away from the Earth to make this cosmic connection possible. Instead, they give the impression that they are deeply connected to the being of the Earth. In worker bees, this connection is expressed by turning toward tangible streams of substance, but drones are oriented toward the Earth's spiritual being. Drones are carriers of cosmic impulses and connect the colony to the spiritual being of the Earth. This is why in their nature drones seem largely unimpressed by their surrounding physical circumstances. They even need to be fed and are considered lazy and sluggish. But they could not accomplish their task for the bee colony if they were of a different character.

Through the collective of drones, honeybees are able to open up to effects coming from cosmic expanses. These form the cosmic chalice of a bee colony. The most peripheral spiritual sheath is given to the colony by the drones.

One key difference between worker bees and drones is that drones only remain with the colony until late summer. As soon as they are no longer needed to mate with a queen, they are driven out of the hive by the workers and starve. During the time following, the colony consists of worker bees and the queen. In winter it enters a special state. It closes itself off from the world of the physical Earth and forms a world of its own. Suprasensible observation reveals that during its winter rest, a bee colony is not located within earthly life conditions at all. It is connected to the cosmic events taking place within the Earth in winter through the queen, who has unbound herself from her body. The influence of drones and their cosmic gifts to the colony are not needed in winter.

In the beginning of the new year, drones begin to hatch again. The colony awakens again to earthly life conditions. Through the reappearance of drones, it once again receives the cosmic foundations of life needed to be effective under earthly conditions.

This depiction does not fully illustrate the mystery of drones. Rather, it portrays a first attempt at deciphering the being of drones. Drones still present many riddles to those who strive to understand the being of the honeybee.

THREEFOLD BEE COLONY

A colony consists of three types of honeybees: a queen, workers, and drones. Each of these three bee beings has very specific, purposeful tasks. The unity formed by a colony is founded in each of the three beings fulfilling its tasks.

The queen is the mother of a colony. All bees come from eggs she has laid. The development and expansion of the colony is birthed through her body. The life of the entire hive depends on this birth stream. The queen was created to conceive the cosmic stream of life from which her colony emerges. She conceives her colony. This is why colonies who lose their queen, and cannot create a new one, will die. Through her winter path into the Earth, the queen also connects her colony to streams of Earth forces. Powerful cosmic and earthly currents unite in the honeybee queen and pass from her to the colony. Through her being, the colony is moved into streams of spiritual forces coming from the Earth and the cosmos.

While a queen does not concern herself with the tangible living conditions of the colony, the worker bees' task is precisely to connect the colony to its immediate surroundings. All changes and refinements of substances come through what can be called the collective body of the workers. Large amounts of substances are collected, transformed, and refined by worker bees. Nectar, honey, wax, pollen, and propolis require their attentive and concerted care and treatment. Rearing of brood, nourishing the queen, as well as supervising the climate in the beehive are part of their tasks. They defend the colony, build and renew comb, create new swarms, and gather into a cluster in winter. When one observes the activities of workers in their entirety, one can only marvel at the wisdom guiding a hive. Everybody tending honeybees experiences that worker bees form a living body, consisting of many thousands of individual beings, through which this wisdom speaks and reveals itself.

As previously described, it is the drones' task to gather at drone congregation areas and mate with new queens. Drones are not taken along with the colony into winter hibernation, but are instead driven out of the hive by worker bees in late summer, whereby the drones die.

The honeybee queen is a being through which spiritual light comes into expression. Her being resides in the quality that reveals itself in sense perceptible light, yet remains invisible. The being of a honeybee queen is not only devoted to the outer Sun, upon which life on Earth depends, but also to the inner Sun, the Midnight Sun, the Earth's Sun.

The queen maintains the connection between the worker bees being effective both in the world of external light and in the spiritual Sun of the Earth. Her secret lies in this task. She provides the workers with a life force that they cannot produce themselves and without which they could not perform their tasks. Through their queen, they are touched and permeated by the light of the inner Sun. This is why she is their *Queen*. If one considered the role of a queen to only be reproduction, the term "mother bee" would suffice. However, she is the queen of the honeybees. This expression goes beyond the motherly and points to higher, spiritual realms of the colony.

We have seen how drones form a community that only reveals itself quietly and unobtrusively. But with this, they are very open to cosmic influences. This cosmic orientation of drones helps a colony find peace and harmony with earthly conditions. The cosmic spiritual chalice formed by the drones supplies the colony with a strength of devotion that helps honeybees enter into earthly existence. It is precisely the nature of drones, oriented to the cosmos, that allows the colony to surrender to the Earth. Preventing the queen from dissolving into the Sun during mating is also a part of the drones' nature. They call her back to earthly life. It is the power of their devotion that causes the queen to return.

The worker is considered the quintessential bee. In numerical terms, the colony consists mostly of worker bees. Their outstanding trait is a changeability of their organism and the resulting versatility expressed in various activities that worker bees master in their lifetime. To a high degree, their organism possesses the ability to remain germinal.

Therein lies the foundation of the workers' ability to create numerous substances vital to the beehive. This ability has already been described. The workers' activities with substances, carried out in a sense perceptible manner, also have a spiritual side. They are the prerequisite for their spiritual transformative work. I have described how the astrality of Earth represents an essential reality for the worker bees. They fly through the astral layers of the Earth. The workers who fly out of the hive are directly exposed to what humans impress on the astral sheath of the Earth with their soul impulses. But the bees don't leave that astrality as it is. They transform it. They illuminate it through their flight and activity. This transformative work is the deep mystery of the worker bees. Through the workers, what spreads as cold, gloomy, and frozen astrality over the living world of the Earth, becomes to a certain extent brighter, illumined, and colorful again.

In summary, one could say that each of the three bee beings is, in its own way, effective in forming a sheath for the entire colony. This effect is most obvious in the queen because due to her path in winter, through the night of the Earth, and also through her marriage flight, the colony remains connected to earthly and cosmic conditions. Workers also carry out sheath-forming tasks. They are the most intertwined with the Earth and the physical conditions of life. Their devotion to the substances of Earth, the transformative processes which these substances undergo, and all of their activities, also form a sheath for the beehive. Of all the spiritual sheaths of the hive, this one is closest to the physical world. Worker bees who sweat wax out of their bodies and shape it into comb are the ones that create the inner structure of the colony. The drones provide the colony with an additional sheath through their existential being. As described, it is of a cosmic nature. Through it, the colony is integrated into the cosmos.

The Being of the Great Bee

The Great Bee and the Earth

Every animal species is shepherded by a higher being of the spiritual world. Of course, this also applies to bees. This exalted bee being bears the spiritual archetype from which honeybees are created. The forming and creative forces that cause honeybees to be brought into existence have their origin in the being of the Great Bee. It is the bearer of the wisdom that reveals itself in individual bees, right down to the structure of their bodies, and from which worker bees, queens, and drones emerge.

The sphere of impact of this higher being—which is also called the group soul of honeybees—extends over the entire Earth. Suprasensible observation shows that it has an inclination toward the higher beings from which emerge animals, plants, and minerals. The angels of various kingdoms of nature form a whole. This means that the angels who guard the archetypes of plants and animals stand in very specific relationships to one another. Life streams of great importance flow between them. Without these streams of relationships, no being of the spiritual world could exist. This applies equally to the Great Bee, as to the other beings who shepherd the archetypes of minerals, plants, animals, and also humankind.

As I mentioned, the Great Bee, the group soul of honeybees, is deeply connected to the Earth itself. It has a very meaningful connection to the inner Sun of the Earth. The Great Bee is united with the Earth as the bearer of the sacred, renewed, rejuvenated Earth. The new, redeemed, future Earth already lives in the Great Bee. This is why it is a being that extends through and over the entire Earth. The Great Bee is not limited to just bee colonies, but has tasks that include the entire Earth and its evolution. The Great Bee is an exalted being among the animals. Its extensive tasks do not end with the preservation and creation of bees. It is devoted to the Midnight Sun of the Earth. The power and essence of the spiritual Sun of the inner Earth lives in the archetype of bees. Bees are alive through it and for it. It is their true mother. The Great Bee is in a sense the mediator between an individual bee and the Midnight Sun of the Earth.

Christ entered the Earth through the events on Golgotha at the turning point of time. He has become the High Spirit of the Earth. The Great Bee is loyal to Him. But the individual bees also serve Him. They are the children of the honeybee queen, but they are also the children of the Great Bee, for it lives in them and through them, and the bees live through it. And, they are also servants of the Christ Being. The Great Bee stands in the life stream of Christ. And every single bee stands in this stream of life.

The new, redeemed, rejuvenated Earth already lives within the Great Bee. The inner Earth is healed. The intact, healthy, vital core and germ of the Earth comes into appearance in the Great Bee. The Great Bee and all her children, and even the elemental beings connected to bees, all stand in the life stream of the healing and healed Earth.

The task of bees is to rejuvenate the Earth. The being of bees can be understood by observing that they are a living picture of the rejuvenation process of the Earth. The transformation of darkness into light and likewise the transformation of light into darkness takes place in them and through them. Heaviness transforms into lightness and lightness into heaviness, despair into hope and hope into despair. Honeybees are animals of change, of transformation. They belong to darkness and light, heaviness and lightness, despair and hope. The great polarities of life meet within them. They stand between them, exposed to them, and mediate between them with the task of harmonizing the contrasts of life. The Great Bee is the bearer of the archetype of the transformative tasks of honeybees.

Substances of Honeybees

The impulses emanating from the Great Bee can be observed all the way down to the processes involved in the substances honeybees produce. These are impulses of transformation for the Earth. Wax and honey, in particular, are substances related to the transformation of Earth and its rejuvenation and redemption. One cannot look at these substances without simultaneously immersing oneself in the light of the Earth and connecting with it. Earth light appears in them. Above all, they are substances of the inner light of the Earth.

A redemption process can already be observed in the process of wax production by individual bees. Not only forces and effects of light live in the white, virgin wax. In its creation, we can also recognize how an individual bee has the power to extract the wax from the darkness extending over the Earth, the untransformed astrality of human beings. In the virgin wax, a process of transformation becomes visible, which deeply affects human beings. The Earth astrality is relieved and becomes brighter, lighter, and more colorful. The Earth is freed from the burdens of unpurified human drives, motives, and life styles. The production of wax by honeybees has its effects in this realm.

The honey process is also deeply related to human beings. It has a connection with the healing of human blood. The collection of nectar by worker bees and the ripening of honey in the bees' comb both support a process of purification of certain spiritual substances in the Earth, which in part, cause human blood to become heavy and dark. These are extremely destructive forces. Through the Earth, they have a damaging influence on human blood. These dark spiritual substances pull the blood downward, so that it becomes less available to the evolution and being of humankind. All human beings, because they have physical and living bodies, carry the great transforming events of the Earth within them. Our blood is the organ through which most of this transformation takes place. It is continually weakened by these destructive forces that lie within the spiritual organization of the Earth. The bees' work of transforming foraged flower nectar into precious honey serves to liberate the spiritual organization of Earth from the harmful forces influencing human blood.

The Hive

Archetype of the Beehive

Anyone studying honeybees in depth arrives at a moment of being deeply touched and even overwhelmed by the wisdom that lives in every hive. Each colony consists of many thousands of individual animals. Whosoever watches them at work cannot help but get the impression that they are driven and guided by a mysterious, communal will. Individual bees seem to know exactly what to do without it being apparent how they know. The most diversified life processes are taking place simultaneously in every hive. One only has to think about the many streams of substances that are flowing side by side in a coordinated manner. The gathering of nectar and preparation of honey is a very complicated process in which the entire colony participates. Simultaneously, brood is tended to and fed by many, mostly younger, bees. In addition, it is important to regulate temperature and humidity, to fulfill specific tasks determined by the seasonal cycles of the year, and to swarm. A colony even has the ability to respond to constantly changing external conditions, such as weather, pollen, or a unique habitat of the hive.

A bee hive is a wisdom-filled, living organism. The thousands of individual animals live together in a harmonious state. How is this possible, when each individual animal clearly lacks an overall view? A great number of individual beings are constantly making many decisions so that the colony can respond correctly to the perpetually changing internal and external living conditions. Obviously, a higher level of consciousness is at work in the colony, through which the

A Swarm

many individual animals form a whole. But what ideas can one form of this wise consciousness? How does each animal know what to do? Who is directing it? Are there words that can be used to describe this wisdom? These are essential questions that have certainly been asked many times before. Naturally, they are not easy to answer. I would like to be clear that I am not claiming that the following description provides definitive answers to the questions about the archetype and wisdom of honeybees. However, bringing together indications and various aspects can serve in coming closer to conclusive answers.

When a beekeeper speaks of idiosyncrasies of individual colonies, they are referring to what they experience as special in each of the colonies. Through these experiences they have gotten to know their "voices." I do not want to address the question of whether it is justified to apply the concept of individuality to the characteristics of bee colonies. The differences between individual colonies can be huge and clearly perceptible to the observer. It is difficult to find terms that describe these differences. The question also arises as to what is responsible for these differences. Dealing with these complex questions will be left to a later presentation.

Birth of a Colony: The Swarm

Swarming is extremely important to the development of a bee colony because it leads to its division and is therefore the prelude to the reproduction of bee colonies.

When a colony swarms, some of the bees leave their home beehive forever. They move out. However, this exodus is preceded by a division of the colony within the beehive. The part of the colony remaining behind does not receive a new queen until the old queen has moved out with the swarm. The young queen usually only hatches a few days after the old queen has moved out. The part of the colony connected to the old queen moves out with her to establish a new colony. The swarming part of the hive rises as a whirling cloud and surrounds the apiary for some time. The air is filled with flying and buzzing bees, which offer a captivating spectacle to the observer.

Usually, swarming season falls into the second half of spring. It is triggered by the rearing of new queens. They are raised by the colony in very special cells called queen cells. When they have reached a certain stage of development, the swarming process begins. The old queen is faced with rivalry from the newly maturing queens, which causes the colony to separate.

Worker bees prepare the old queen for the excursion by, as it were, subjecting her to a fasting diet. This enables her to fly. At the time of swarming, the bees remaining behind have not yet hatched a new queen. As soon as she hatches, one of the maturing queens will replace the previous queen. The other developing queens are killed by the worker bees.

The departure of a swarm is heralded by a great unrest among the bees. They fill their stomachs with honey as preparation for the time they will be traveling. The swarm only leaves in sunny weather. Thousands of flying and dancing bees form a large cloud and among them is the queen.

Swarming is a state of emergency for honeybees. They have left their home and have not yet found a new one. One gets the impression that they are being sucked into the air and light. A swarm is a powerful formation. It moves through the air like a living being. The bees are in a somewhat intoxicated state. They surrender to the space, the landscape, the light, yet remain connected to each other.

In a dancing cloud, the swarm moves away from the old beehive. After a while, it gathers where the queen settles and forms a tight cluster. Often this forming swarm cluster lands in a tree. From there the swarm has to find suitable housing.

If the beekeeper succeeds in catching the swarm and moving it into one of their beehive boxes, they will have gained a new colony in their existing apiary.

The swarming bees leave behind brood they tended, provisions they stored, and honeycomb they helped build. By doing so, they lose their protection. But they keep their queen. What now follows is a process that causes swarming bees to create anew what they have given up. They search for a place where they can find shelter; then, they build new honeycombs, fill them with provisions, and there the queen can lay her eggs from which the next generation of bees will emerge. Swarming bees break away from the hive in which they were born and lose the connection with the place where they previously lived. This is the beginning of a process that leads to the formation of a new colony. One phase of life passes into another. The bees leave behind the bonds that previously embraced them. The queen, too, is breaking away from her original colony and the place she has lived until now. A swarm that has left its ancestral hive is not yet a colony because it is not yet capable of survival. The conditions that would ensure survival have not yet been found. A swarm could therefore be considered the germinating seed of a new colony.

While the queen goes with the swarm, the Bienerich of the existing colony remains behind. Nevertheless, the swarming bees are accompanied by a Bienerich. This is a young Bienerich coming from an apprenticeship with the old Bienerich. When a swarm forms, the old Bienerich calls on one of the apprentices to join the swarming bees. The formation of a swarm is therefore also the birth of a new Bienerich.

Suprasensible observation of a swarm shows that it radiates a dynamic brilliance that is continually renewed. A swarm shines brightly. It provides an occasion through which light is born.

When one watches what happens to this luminous glow, one can see that it is very fleeting. It rises quickly and penetrates into a higher, spiritual realm of the Earth, where it connects with the souls of human beings who are in the process of leaving the Earth. The glow that rises from swarming bees shines forth among the human souls saying goodbye to their earthly life. It unites itself with these souls. It provides them with an orientation that can be important for them. Because the dead leave behind their orientation to the Earth, they are now in danger of being seized too quickly by the might of the spirit. This means that the spirit would have a kind of soporific effect on the souls. The stream emanating from bee swarms provides them with a form-giving power that helps the souls find an inner cohesion. One could say that the souls stay more alert as a result. Honeybee swarms radiate an awakening impulse into the souls of the dead.

By dissolving the unity of the colony while swarming, the honeybee being touches the souls of human beings who are in transition between the earthly and spiritual worlds. Honeybees do reveal their being to humankind, but they do not reveal it to those who are living on Earth, rather to those who are between worlds. This is a significant insight because it reveals the connection between bees and humankind. Bees meet human beings in the physical world, but they also meet them in the spiritual world. Bees are beings of transition. They stand between worlds. What their being shows of itself under earthly conditions is only one part. They do their work not only for the Earth, but also for the souls of humankind.

Swarm Cluster

The act of swarming elicits a need among the bees to gather again in one place. Often this happens on a tree near the former beehive. The swarm, consisting of many thousands of individual beings, contracts to form a single cluster. The queen is among them. Michael Weiler describes this process as follows:

> It is now at least five minutes since swarming began. The hum of the swarming bees is creating a uniform sound picture. The number of bees flying to the swarm cluster is increasing and it is slowly getting bigger. . . . About fifteen minutes from the start, almost all of the bees are in the swarm cluster, which has now grown to roughly the size of a rugby ball. A swarm of this size weighs around 2.5 kg (5.5 lbs) and contains about 18,000–20,000 bees along with the stores of honey they consumed before swarming. There are also drones in the swarm and bees with pollen baskets, which have joined the swarm after returning from their foraging flight. After twenty minutes, peace returns once more to the apiary; bee traffic is as before. From a short distance away, the swarm seems to hang motionless in the tree. Most likely a casual passer-by would not even notice it.[4]

One can easily sense that with the formation of the swarm cluster, a great and festive event is occurring. To stand before a swarm cluster is an amazing experience. An almost indescribable magic and a deep, mysterious calm emanate from it. One feels deeply moved and does not even know why.

When the bees gather to form a cluster, they are still on a journey. They have not yet found a permanent home. Scout bees set off and search the area for suitable places where the new colony could settle.

My research revealed that something very significant occurs when swarming bees first gather together in a cluster. It is the moment in which the swarm, which has separated from the original colony, finds itself. With this an essential phase is reached, which will eventually lead to the swarm becoming a colony of its own. The calm that one experiences around a swarm cluster is the sensible expression of this process.

When a swarm flies out, the bees break away from the influence of the being that we have come to know as the master elemental, the guardian of the apiary. It has no influence on the swarm. The swarming bees slip out of its cloak. In doing so, they can accomplish their task of finding themselves. Then, when honeybees form a swarm cluster, they return into its cloak. The resting swarm reposes again in this being.

The formation of the swarm cluster is of great importance to the young Bienerich. After it has gone with the swarming bees, it will take on its new vocation. It can be observed that the new colony is entrusted into the protection of a young Bienerich as soon as the swarm cluster has formed. This is a meaningful moment. The Bienerich connects with the swarmed queen. It puts itself in her service. If one observes the Bienerich at this moment, one can see its excitement to be among the bees. It knows the importance of this moment and senses the honor that comes with it.

It can be observed that at this moment something takes place that can only be described as a sacred act, a ceremony. It may be surprising that such rituals exist in a world not perceptible by the senses. But whoever feels or looks into the spiritual reality of Earth will always have the opportunity to witness such sacred acts. Elemental beings are part of these ceremonies, which are actually performed by the beings of higher spiritual hierarchies. There are a number of rituals among the elementals, which even include processions, a kind of religious parade. Special moods of nature—for example, the rising of the Sun—are indications that invisible rituals are taking place in the elemental and spiritual worlds.

Spiritual observation of these events confirms that when a swarm cluster forms for the first time, the spiritual archetype of the colony is present among the swarming bees. The sacred order of a bee colony flows as a living force between the bees. Through this process, the new colony is initiated.

The bees forming the cluster are permeated by the primal image of the bee colony in a way that is difficult to describe. The archetype flows through the collected bees, so to speak. The colony arises when the swarm of resting bees unites itself with the task given to honeybees by the spiritual world. By listening to what happens with and within the swarm cluster, one can approach the task that the bees have received from the spiritual world. At that moment it rings out, spoken by the spiritual world. A new colony is created through what is spoken and what resounds.

The presence of higher beings of the spiritual world can be perceived during this process. The formation of a bee colony is an event also attended by the spiritual beings of the landscape. In this moment, time stands still.

Moving In

After a certain time, sometimes only after a few days, the swarm cluster dissolves. Michael Weiler describes this process and what happens afterwards.

> Suddenly the behavior of the swarm changes. Vibrating their wings, the bees run around very fast seemingly in a muddle. Things start to loosen up, the cluster starts to disperse. In a short time, the bees leave the place they started from and fly away, all in the same direction, in an elongated cloud. If they do not fly too far, the observer may be able to follow them and can watch how they gather at a new place and there disappear, possibly through a

hole or a crack, into a cavity. This could be a hollow tree or a crack in a rock; but it could also be the false ceiling of a half-timbered house or a cavity wall. Other possibilities include large empty nest-boxes, chimneys, and, not least, an unoccupied hive left open at another apiary.[5]

He describes how the swarming process that leads to the formation of a new colony ends when a colony moves into a new home. When this happens, the bees immediately begin to build new comb, which forms the physical foundation of their life together.

After the initial comb has been built, the queen begins her task of laying eggs. The continued existence of the colony is now assured. The process of its renewal, which began with the division of the original colony and the subsequent swarming, is now complete.

Twelve Aspects of a Bee Colony

In the wisdom-filled interplay of bees, certain laws emerge that cannot be explained if one acknowledges only a physical reality. Bees demonstrate that the concepts of the physical world are insufficient in explaining their coexistence as one colony. The creative principle of Earth is of a spiritual nature. It is self-evident that the spiritual archetypes of life are not sense perceptible. But even if our sense organs are not able to grasp spiritual life principles, these are nonetheless expressed in living beings. They reveal themselves in the appearances manifesting in individual beings. These appearances are accessible to the senses. The path to recognizing the spiritual archetype of a being begins with a study of the individual expressions of life in which the archetype reveals itself in the earthly realm. By studying particular aspects of a being's life, we create the possibility of approaching not only the being as it reveals itself to the senses, but also the spiritual archetype from which it originates.

On the one hand, spiritual archetypes remain hidden behind individual phenomena; they are imperceptible to the senses. On the other hand, one can grasp the spiritual form of a being by immersing oneself in its life phenomena. A lively study of the details leads to the whole, to the archetype, if one adds an inner, soul activity to the sensory perceptions. The sense perceptible details are read like revelations of the archetype. An archetype has a spiritual nature. What the senses show of things and beings, however, is a medium, which human beings can read by way of their inner soul activity. The more precise and clear the sensory observation, the easier it becomes to advance into the realms of spiritual life from which the sensible emerges. It is essential that one can accept individual phenomena or aspects of life as revelations of the underlying spiritual essence.

Now, we will begin to tread this path characterized above in order to understand the archetype of the bee colony. If one delves into the expressions of the life of a bee colony, one will find twelve aspects through which the entire being of honeybees can be grasped.

Every bee colony goes through an annual cycle, the main seven stages of which can be described as swarm, swarm cluster, moving in, day-night rhythm, marriage flight, overwintering, and winter cluster. There are also five aspects that describe other important characteristics of a bee colony: its relationship to humans (beekeepers), its relationship to the landscape, the bees' healing remedies, the light of honeybees, and the death of a bee colony.

Each of these aspects has an important meaning for the life of a bee colony. Through suprasensible observation, one becomes aware of the deeper significance of each of these aspects. The results of such research are briefly presented below.

The rebirth of a colony begins with the creation and flight of a swarm. The old queen leaves the hive and moves out with part of her colony. The forming *swarm cloud* becomes the focus of a powerful spiritual event. Creative forces to which a bee colony owes its existence are present in the periphery of such a cloud. A swarm of bees is surrounded by the forces that have always accompanied honeybees all along their way, down to the form in which they present themselves to the senses today. Since these forces have been accompanying Earth's evolution for a very long time, the path of Earth from its primeval beginnings to the present day is perceptible around a swarm. It is even possible to see the coming, future stages of the Earth's development. A swarm of bees is an earthly event in which Earth's evolution becomes spiritually visible. The laws of development, the stages of transformation, and the forces associated with them appear to the clairvoyant.

After a certain time, the swarm collects and condenses to form a *swarm cluster*. When observing a swarm cluster, the clairvoyant or clairsentient perception is drawn into the inner spiritual structure of the Earth. The Earth appears as a light-filled, living being. Each swarm cluster forms a gateway that connects the depths of the Earth to the surface of the Earth, meaning the spiritual Earth with the sensory Earth. It is as if a kind of breathing occurs through this gateway. Messages to humankind move through the bees' swarm clusters. Just as a swarm cloud makes visible the creative forces of Earth, the swarm cluster resounds with words addressed to humankind, telling them of the significance of the Earth as a spiritual being.

The bee colony eventually moves into a new home, often made available to it by the beekeeper. *Moving in* is a great moment for the colony because the time of uncertainty is over. Contemplation of this process particularly touches the soul. One is witnessing the birth of a being that has previously existed in a spiritual state. The colony now has an earthly home. The human beings who make such moving in possible are touched at this moment by the spiritual being of honeybees. They are being thanked deep within their soul. They are the ones who ensured that the colony survives. It is due to their attention and care that the homeless colony has been given an abode. Moving in is therefore a moment of great significance for bees as well as beekeepers.

Once the colony has a home, it can begin to settle in. It builds honeycomb, the queen lays eggs, and the surrounding flowers are visited. The colony begins to develop its activity. The *rhythm between day and night* has a special meaning for the activity of a bee colony. Without this rhythm, the colony would not be able to develop in earthly living conditions. The night heals the wounds of the day, and the day reveals the events of the night. What is visible of bee activity is only an outer cover of the mysterious healing work that honeybees perform for humankind and the Earth organism. The bees' diligent work on the substances they collect and transform points to an invisible spiritual event, by way of which the bees have a healing effect. The honeybees' secret is the effectiveness of their healing.

A virgin queen must first mate with drones to be able to start laying eggs. To do this, she rises up into her *marriage flight*, during which she mates with a number of drones high in the light-permeated air. The queen's flight leads her to unite with the light of the Sun. She burns up spiritually in the light that shines down onto Earth. But actually, she is reborn through the Sun, and it is by this that she is the queen of a bee colony.

When the days get noticeably shorter, the colony begins to prepare for winter. The last bees hatch and the colony contracts. This is called *overwintering*. During the cold season, the colony is nourished by the provisions they gathered over the course of the year. The amount of honey plays a significant role in the future of the colony. Will it last until spring? Summer is omnipresent in the substance of honey and it continues to work in it, enveloping the colony in warmth and light. Honey protects and nourishes the colony and gives hope that a time will come when they can fly out again.

It is winter. The bees are in their *winter cluster*. They have contracted around their honey supply. The queen unites herself with the light of the Earth. She does this for her bees who protect her body from getting cold. While the winter bees are guarding the queen's body, she can gather life substance for the colony from the spiritual depths of the Earth. This is the only way the colony is able to fulfill the high tasks assigned to it.

The next aspect applies to the *individual relationship between beekeepers and their colonies*. According to their inner natures, a deep spiritual relationship exists between human beings and the bee colonies they tend. The spiritual nature of the bees touches the beekeeper, even if he does not notice it. Their soul is the carrier of the being of bees, which means that this being participates in shaping the beekeeper's soul, down in its depths. On the other hand, the beekeeper takes on a task that affects her life significantly. The bees, their needs and their rhythms, determine the beekeeper's everyday life. In this way they help the bees live, to be present on Earth, and to fulfill their tasks. Beekeepers are by all means constantly active as midwives of the bees.

The *landscape* extending around the beehive is of extraordinary importance to the colony because bees feed on what the plant world produces. But even the qualities of the various elements—earth, water, air, light, and warmth—are vital factors to the existence and development of a bee colony. A colony is always an expression of the forces active in the surrounding landscape.

Bees not only create honey, which is used as food and medicine, but also other substances that serve exclusively as *remedies.* These include propolis, wax, foraged blossom pollen, and also bee venom.

Another aspect that deserves attention is *candlelight.* The light that emanates from burning beeswax candles expresses the essence of honeybees in a special liveliness and form.

The *death* of a bee colony is an unforgettable and devastating event. But there is no death that is not a new birth. Death is always *transformation.* This relationship between life and death is the theme of the last, the twelfth aspect.

In autumn of 2011, I was able to define these twelve aspects and receive a verse for each of them. These twelve verses are written in a later chapter (Chapter XII). They can serve in finding a meditative approach to the various aspects. Each one reveals a part of the spiritual archetype of the bee colony.

Essence of Honey

Honey is an expression of the activity of a bee colony. A colony creates honey, but it is also the foundation of all the hive's liveliness and activity. Bees could not do anything without honey. It nourishes the colony in a way that enables the workers, among other things, to sweat out the wax for building comb. Thereby, honey is also the foundation of the honeycomb structure, which offers the colony the opportunity to exist permanently in the physical realm.

One could say that the being of the colony reveals itself in the honey at the level of substance. The being of the hive can be experienced in its honey. In the substance it produces, the archetype of the bee colony unveils itself. This is why we will discuss it here.

The nectar that the workers collect from flowers is the foundational substance of honey. The being of each respective plant is inscribed in the nectar, which emerges in droplets from the calyx. The nectar represents an essence of the plant; it is a plant that has become substance. In the blossom, the plant being appears as an image, in the nectar it becomes a substance.

But the nectar expresses not only the plant being; it also reveals the character and being of the landscape in which the plant is growing. The nectar contains traces of the spiritual constitution of that piece of Earth to which the corresponding plant belongs. The spirituality of the Earth is inscribed in nectar. The water processes, the air and light processes, and the warming processes of the landscape move through each plant. It can only become visible, because in each plant the mysterious working and weaving of the spirits of the elements condense into a living expression. Every plant works differently with the various processes of the elements. In the substance of the nectar, we can find a spiritual trace of events involving all the elements in the landscape. This trace is retained during the purification of the honey. It ends up in the honey almost entirely unaltered. The selfless nature of the worker bee guarantees that this delicate imprint of the landscape remains in the honey. Nectar brought into the hive is exchanged between worker bees and thus flows through the entire colony. This gives it a texture that enables it to be stored in the honeycomb.

A colony is nourished by the stored honey. It is its food reserve. This is especially true during the winter time of rest when flowers no longer supply a source of food. The honey the bees have created enables them to survive the cold season.

In the honey collected by a colony, one can also see the memory of the past months condensed into substance. During hibernation, when the colony is feeding on their honey, they are ingesting this memory as food. The memories of summer nourish a bee colony in wintertime.

The healing effect of honey on the human organism is largely due to the fact that the concrete events occurring in the landscape during the bees' foraging have been spiritually preserved in the honey. Honey is a substance with the special property of being able to absorb the life of the elements of Earth like a memory. The pure nature of bees is expressed in the fact that they do not take away from the honey what the landscape has imprinted on it.

It is important to note that honey has a fairly immediate effect on human blood. Suprasensible observation shows a deep connection between honey and blood, a mysterious attraction and kinship. It seems as if honey strives to pass its effects on to the blood quickly. In a certain sense, honey has a similar relationship to the landscape as blood does to the human organism. As a substance, blood is an essence of all physical processes. One can basically determine what is happening in the body by looking at the blood. In a similar way, honey is a living mirror of the events of the outer world, as the blood is a mirror of the events of the human inner world. It is precisely due to this relationship that honey has a healing effect on blood. Honey combines with blood because—mediated by the work and nature of bees—it is to the landscape what blood is to the human organism.

At a certain point, the honeycombs are covered by the bees, sealing the honey-filled cells with a thin layer of wax. Beekeepers have to scrape away this layer if they want to harvest the honey. Only then can the honey be extracted.

By being sealed in the comb cells, honey becomes receptive to the action of forces that rise from the Earth into the beehive. During the curing process, Earth light shines into the honey. The shape of the honeycomb cells attracts and concentrates this light of the Earth. The strong formative forces in the light make their way into the honey substance. They are able to bring any unbalanced life forces in human beings back to a healthy equilibrium. Through the honey, creative forces permeate the organism and help it to regain its own intrinsic order.

The healing power of honey in human beings is not only due to the aforementioned influences of the landscape, but also to light processes that take effect in the honeycomb during the ripening process. The radiant light of Earth connects with what lives in honey as a memory of the events of the elements.

Honey is a substance in which many influences unite to form a whole. The special formative force of a bee colony is materialized in honey. The being of the bee lives in the honey process. The mildness, benevolence, and harmony inherent in honey are bee forces. They are an expression of the nature of honeybees. By consuming honey, human beings allow their organism to participate in the creative power of a bee colony.

Inanna Frey has been receiving messages from bees for many years. She describes the essence of honey with the following words:

> The being of humankind corresponds to the essence of honey.
>
> Thus is it one with the being of the Earth.

Bee Venom

Rudolf Steiner described how bee venom plays a significant role in a colony's tendency to swarm.

> When a young queen hatches in the beehive, then, as I have told you, there is something that begins to disturb the bees. Previously, the bees were living in a kind of twilight. Then, they see this young queen light up. What is connected with the illumination of this young queen? When the young queen lights up,

she takes away the power of the bee venom of the old queen. The fear of the departing swarm is that it no longer has bee venom, can no longer defend or save itself; so, it moves out.[6]

This observation by Rudolf Steiner gives us reason, when contemplating bee venom, to concentrate on more than just the effect of somebody being stung. Bee venom also has a significant effect on the bees themselves, especially since it is produced by the bee organism. Every single worker bee is affected. Suprasensible perception shows that the effect of bee venom is particularly important for the formation and unity of a colony. A bee colony could never develop without bee venom. The poison generated in its own body enables the individual bee to subordinate itself to the unity of the colony. Bee venom paralyzes the tendency of the individual animal to express its uniqueness. It suppresses the individual bee's separateness in favor of the colony as a whole. Only through the venom's binding effect can the sum of many individual animals become one single colony. In this context, it is worthy to note that drones, which one could experience as foreigners in the bee colony, do not produce any venom.

Bee venom is the carrier of a very specific astral force, which leads to each single creature being deprived of the opportunity to develop an individuality. In this way it can open up to and realize supra-individual impulses. The venom paralyzes the peculiarities of individuals in favor of a higher principle.

It is interesting that people's respect for and fear of bees is due to their venom. A bee sting is painful, and its consequences can be very unpleasant for humans. That which creates the opportunity for a colony to form also protects it and provides a necessary respect.

Honeycomb

We must not forget the honeycomb structure when talking about the forces that hold a bee colony together. It is of outstanding importance to the colony. Worker bees create it from small wax scales that they sweat out of their abdomens. In his book, Michael Weiler describes it like this:

> The bees link together in a sheet by forming chains and hang from the lid or some other upper boundary (top bar of a comb frame, etc.) in the cavity that they occupy. Increased heat production can be detected in the construction cluster (greater than 36°C or 97°F) as a number of bees exude wax scales. These are carried upward with the legs and mandibles. The bees at the top of the construction cluster knead the wax with their mandibles, while at the same time probably mixing it with a secretion from one of their cephalic or head glands. Then they attach it to the frame bar from below. The wax is first drawn into blobs and a little later is shaped into a hexagonal matrix. By applying more wax to the edges of the developing comb its area is increased. As soon as this has reached a diameter such that, if rotated around its vertical axis, it would sweep out a sphere of about 5–7 cm (2–3 in), they start the construction of adjacent combs. This way the comb matrix grows until eventually it hangs in the space occupied by the swarm.[7]

One can admire the precision with which honeybees build comb. If one looks at the significance of the honeycomb structure to the colony, one will find that it represents the "Earth" for the bees. Honeycomb is the physical foundation of a colony. And what is remarkable is that the bees create it themselves. Honeybees create the very Earth they live on!

We can only understand an individual bee and a whole colony if we consider their connection to the honeycomb structure. Without the comb, the colony is inconceivable. It is an essential part of the being of the bee. By building comb, honeybees bring forth something out of themselves that is close to and inherent in their nature. The comb structure forms their physical foundation for life. Suprasensibly, one can see that the comb structure is of outstanding significance to the colony because, to the bees, it is not rigid the way it appears to us. To them, the honeycomb structure is anything but rigid. It is highly flexible, and, one could say, sounding and resounding. Honeybees hurry across the comb as if running on something that is rather fluid to them, something that moves and connects them to one another, but also resonates with the world outside of the beehive. Bees can only accomplish their transformational tasks because the wax comb creates for them a resonating motion. As a result, the bees of a colony have something in common and something that connects them. Honeycomb creates something that can be described as a colony's resonating center. At the same time, the comb is the place where brood, the future of the colony, develop. Pollen and honey are also stored in the wax comb as provisions for winter.

We can now explore another context, which is difficult to describe, but should nonetheless be approached. Honeycomb gives the bee colony a holy center, and therefore creates not only an essential support for its tasks, but also a bond with human beings. Viewed imaginatively, honeycomb envelopes the organ of the human heart. That may sound like a strange idea. For the bee colony, however, this connection is real. The comb structure, shaped of many wax cells, forms a connection to the human heart. The colony feels through the honeycomb structure what can be called the heart region of the human being. The comb appears to the bees like a chalice, through which they discover what human beings carry in their hearts. Honeycomb is a suprasensory gateway between the human heart and honeybees.

Warmth

The honeybees' ability to produce warmth is one of the characteristics that sets them apart from all other insects. A worker bee's organism is able to generate heat that benefits not only the individual animal, but the whole colony. A bee colony is constantly regulating its own warmth. Its survival depends largely on whether it is able to maintain the necessary temperature. In this context it is very interesting to think about how a swarm deals with this. A swarm does not yet have warmth of its own. It is completely dependent on the warmth of the air through which it moves. A particular temperature is therefore an absolute prerequisite for a bee colony to swarm. It is in the forming swarm cluster, especially the moment the new colony begins to produce wax and build honeycomb, that a special warmth begins to arise. To achieve this, the muscles in the thoracic region of an individual bee, which are used to move the wings, begin to vibrate. However, the wings are "unhooked," so that they remain still and the muscles otherwise used for flying can now generate heat. Although each individual animal can produce heat, only the unity of the colony is able to hold and regulate this warmth. Working with warmth is a collaborative effort.

From a suprasensible point of view, one could describe warmth as having the task of connecting the spiritual archetype of the bee colony to the individually manifested colony and its activity. Warmth attracts the spiritual forces that cause the collection of bees of each colony to submit to a higher principle, the archetype of the bee colony. Without the warmth the colony produces, the spiritual archetype of the colony could not become effective among honeybees. Warmth shapes the bees into an organ of this archetype.

Spiritual Earth

By directing our attention to the place from which the archetype of the bee colony originates, we become aware of an inherent mystery of honeybees. This is certainly not easy to do. It can be noted, however, that the archetype of the bee colony emerges from the Earth as a special force. The connectedness of honeybees, their ability to form a colony from many thousands of individual animals, is a gift from the innermost Earth. Bees are connected to the spiritual Earth insofar as they are a united colony. This observation may come as a surprise, but the importance of the Earth for bees has already been described several times. The honeybee queen falls asleep in winter and connects with the spiritual layers of the Earth. However, she does not walk this winter path for herself, but for her colony, which can thereby incorporate crucial life impulses. If one descends into the spiritual layers of the Earth, one encounters a sphere of life in which the archetype of the bee colony can be found.

Bee Being among Transformative Processes of the Animal Realm

Transformative Work of Bees

Transformation of the Earth

Higher elemental beings have repeatedly made themselves available to me in my research. They have answered my questions and pointed me to possible solutions. These exalted elementals exude great dignity and wisdom. They emerge in one's awareness in such a way that one feels deeply touched. I have often experienced that they wish to work into human consciousness in a way that allows them to express their concerns. It is only possible to encounter them if the impulse comes from them. If they would not allow us to notice them, they would remain unrecognized.

One being that has become very important to me during my research of bees belongs to this group. I have learned a lot from this being about the spiritual nature of bees. It is a being that calls itself the Guardian of the Earth. It does not live exclusively in a certain location on Earth, but is able to move quite freely. When it appears to the observing human being, it radiates loving sympathy. One gets an unmistakable impression that it knows much more about us than we know about ourselves. This is a special experience. But one never gets the feeling of being exposed. A loving and very peaceful mood emanates from this Guardian of the Earth, which does not cause us to be ashamed or fearful in the presence of this higher consciousness.

The Guardian is connected with the being of the whole Earth. It is a carrier of the spiritual life impulses of Earth. The Guardian can therefore live in all of its conditions, knows its layers and regions, and can extend into the vastness and depths of the Earth. It visits humankind as it sees fit. However, one can call on the Guardian and ask questions. More than any other, this being has helped me to understand the honeybees' transformative work on Earth.

The Guardian of the Earth has taught me to look at bee colonies in a certain way. It taught me to look at bees *through* the Earth's perspective. This viewpoint is unusual. But it is possible to look at bees as part of the Earth planet. To do this, one can move into the ground and from underground, meaning, from below, direct one's inner gaze at a colony of bees. In doing this, one becomes aware of the connections alluded to. This change of perspective leads to intimations concerning the contribution of bees to the well-being and prosperity of the entire Earth.

The indications I have received from the Guardian of the Earth illustrate a way we can view the relationship between bees and the Earth. Bees are not to be viewed as separate from the spiritual organism of the Earth. They are beings of the Earth. They come from the Earth. Their life cannot be separated from that of Earth. Every bee colony has a very powerful relationship with the Earth planet in its entirety. By following the instructions of the Earth Guardian, one becomes aware of the fundamental difference between bee colonies in summer and bee colonies in winter, in respect to their relationship to the Earth.

Suprasensibly, only a very limited separation between a colony and the Earth can be observed in winter. The bee colony connects with the inner spiritual light of the Earth. Every bee colony reposes into the Earth in winter, so to speak. There is a strong mutual saturation between the colony and the spiritual Earth. The colony dissolves energetically into the Earth. It loses its individuality, but it gains the light of Earth.

Observing this suprasensible event, one gets the impression that this is very important for the bee colony. It changes into a state that can be described as its primal state. It plunges into the world from which it has come. It rests in its origin. One cannot help but realize that the nature of a bee colony can only be understood by really understanding this state. Wisdom permeates the bee colony and shows itself in every detail of the bees' coexistence. We come nearest to this fact by observing what happens spiritually when bees form a unity with the Earth in winter.

During the light half of the year, in spring and summer, things are completely different. The colony is active in the landscape. Its spiritual relationship to the Earth is such that one could say the identity between Earth and the colony is only slightly developed. Colony and Earth do not permeate each another, but rather the colony is living as an independent entity in relation to the spiritual Earth. It has emancipated itself from the Earth to which it belonged in winter and will belong again. This movement, this release from the influences of the spiritual Earth, is the prerequisite for what can be viewed as the transformative work of honeybees on the Earth. In summer, the colony stands apart from the spiritual Earth. It has acquired a certain degree of independence. It is precisely this independence that enables the bee colony to do the work of transformation that we are talking about here.

By suprasensibly looking at what is underneath our feet, we can see that the Earth is inhabited by an enormous number of spiritual beings. Even in the apparently lifeless minerals, we can find soul and spirit life. There is no earthly substance that does not also have a life of soul and spirit. To suprasensible observation, the Earth is alive down to its deepest depths. It is constructed in such a way that it consists of spiritual layers that can be observed by the clairvoyant. One can study these in a similar way to the layers of the atmosphere that are perceptible to the senses. Each of these subterranean spiritual layers is inhabited by spiritual beings. In this context, the spiritual structure of the Earth will not be discussed in more detail. But I should mention that these beings are of highly varied nature and type. One can encounter high beings serving the general progress of the Earth. But there are also those who inhibit, lame, and fight growth and evolution. These inhibiting forces work far into the human kingdom. Yet, to a certain extent, human evolution is inconceivable without these inhibiting beings. Many human capacities exist by having access to these forces. But we must not allow ourselves to be dominated by them. When that happens, these forces turn into obstacles that are difficult to overcome and they become "evil." These spiritual beings are deeply connected to the Earth and are very strong. They are working to dissolve the evolution of Earth. They are imbued with an incredible power of separateness. They are selfish beings who, by sending impulses through human beings, gain strength and influence. They want two things: on the one hand, they want to alienate human beings from the Earth by making us selfish; on the other hand, they feed the need in humans to want to dominate the Earth by tempting us to act ruthlessly and follow ideas opposing life.

But these forces also serve human beings. Only through them is it possible for us to understand ourselves as individuals and thus to strive to achieve inner freedom in our being. Only by seizing and changing the Earth can we take responsibility for it. However, this must not lead to allowing ourselves to be dominated by these forces. Our task is to find a healthy, sacred balance in dealing with them.

Next to these forces of resistance and negation, there are also beings active in the Earth that support its development. They fully stand in the stream of the evolution of Earth, as it is intended by the spiritual world. There is an intended evolution of Earth and humanity. It is of great importance that humans connect with these beings because the awareness of their activity helps support us in having a living and real relationship with the Earth. The religious, spiritual life of human beings is ultimately the way to connect with these beings.

Spiritual observation shows that honeybees are not only beings of light and warmth. They also possess a surprising tendency to associate with the darkness and cold of the Earth organism. Honeybees live in the boundary, on the threshold between light and darkness. If all we see in them are sisters of the light, then we only grasp a portion of their being. Their importance is revealed in the fact that they are drawn towards both darkness and light. Their sisterhood with darkness is just not immediately obvious. In order to fully understand them, one must lift the veil behind which their dark being is hidden.

As previously described, bees tend to the forces of darkness and destructiveness that live in the Earth without being absorbed by them. They are indeed sisters of light, actually of warmth, but they do not flee the darkness, rather they expose themselves to it without becoming dark themselves. It is really very moving to see how close bees come to darkness, to the destructive forces, without losing their own innate light.

We can also observe that these forces are transformed by the activity of bees. Honeybees have a transforming effect on that which is cast down by the work of the beings of paralysis and resistance in the spiritual layers of the Earth. Through their existence on Earth, that is, in the visible world, bees have the ability to lift up what makes the Earth heavy and polluted. Without this hidden spiritual work of bees, which takes place in the background of their visible work, the Earth would not be what it is now. It would be much drier, much harder. The plasticity of the Earth, its changeability, the tenderness needed for processes such as, for example, seed formation, would be much less developed. Honeybees affect the Earth's ability to recover and free itself again and again from the observable effects of adversarial beings.

By investigating this transformative work more closely, one can make a very surprising discovery. A special current can be felt that rises from the Earth and flows over the flying bees. This stream of renewal is connected to the activity of flying. It rises from the Earth and is released through the rapidly moving wings of every flying bee. In this way, the ascending stream flows through the colony and its queen. At the tips of the bees' wings, it is again released. The collective movement of the wings of a whole colony forms a suprasensible sphere. This sphere creates a type of suction, by means of which transformation is brought about. This is the driving force behind their transformative work. The need for this transformation lies in the Earth itself, and is a need of the higher and lower spiritual beings bound to the inner Earth's existence. Honeybees provide these beings the opportunity to free themselves to a certain extent from the negative forces that inhibit their evolution.

These descriptions shed a meaningful light on the essential forces connected to a bee colony. In fact, great transformative forces become effective through honeybees and reveal themselves when one looks closely at how deeply bees are engaged with the negative forces in the Earth and how near they come to them. Here, we become aware of a significant mystery of honeybees. It touches a sphere of life on Earth and in humankind belonging to the most holy aspects one can encounter on the path of suprasensible research: the effect of the Cosmic Christ Being in the Earth. My research has repeatedly confirmed that honeybees reside entirely in the Being of Christ. One cannot really understand them in their totality if one does not take their connection to the Christ Being into consideration. I believe that many people of the past suspected or even recognized that honeybees and the Christ Being are in a deep and purposeful relationship with one another. Even today I know many people who have no doubt about this connection. This is very encouraging.

Here, I would like to add the words of a seasoned Bienerich.

> The mother of the sister bees is their queen; their father is Christ Himself. They are like His children. The task of drones is to convey His paternity. They pass on His being to the bee colony. They are therefore entirely in His stream, because only through them can He transfer his fatherhood. Drones are exactly who they are because this transmission is their essential task. Honeybees reside entirely in His grace, in His protection. They can only achieve what they achieve because they are actually part of His Exalted Being, in a way that very few beings on Earth are. There is only one comparable being on Earth—the human soul.

Human Sleep

In Chapter V, in relation to the gifts of worker bees, we described another transformative task that bees carry out on sleeping human beings. I will therefore only briefly mention it here again. An emanation can be observed suprasensibly, which rises from active worker bees into the sphere of sleeping humans. This emanation makes it easier for the sleeping soul to find a balance to the physical living conditions with which it was consciously connected when awake. The sleeping soul is given the opportunity to put aside all that which still binds it too strongly to daily experiences, and to let them go. This letting go, this farewell, is what guides the soul to surrender to sleep. The ability to let go is a grace for the soul, through which the soul comes to itself. Through this grace, the soul is touched by the same reality that later it will experience upon death. Sleep is a little death. The work of the honeybees, collecting and transforming certain substances, deeply intervening in the processes of substance on the Earth, releases a healing stream, which rises to the souls of sleeping human beings. Because honeybees bind themselves to Earth substances, forces are set free that help souls to let go of day-to-day experiences.

Realm of the Dead

The third transformative task of honeybees has also already been described. That which arises like a glow from swarming bees unites itself with the realm of the souls of the dead. This enables these souls to share in a force of a very special nature. They look into a life rising up to them from Earth. This is strengthening for them in a very specific way. They are given forces enabling them to connect harmoniously with the world in which they lived just before entering the spiritual realm. Within this world, they receive from the Earth the light that rises to them from swarming honeybees. The dead look to Earth through what rises to them from swarming bee colonies. But they don't look *at* the Earth, they look *into* it. The Earth unveils itself and reveals its inner luminosity. The Earth appears to these souls as the home to which they will return after their journey through the spiritual world. This strengthens the impulse to remove the separation between the physical and the spiritual worlds. Ascending souls, gazing at the light emanating into their realm from swarming honeybees, are permeated by a force that makes it easier to reconcile themselves with their current state. Souls who quarrel with their fate, because they had to leave so much unresolved behind on Earth, are now surrounded by a mild light. The dead are partially freed from their pain because they experience that the separation from the earthly world is easier than they thought. They realize that the earthly and spiritual worlds deeply permeate each other. There is not only a here and there. There is also a here in the there and a there in the here.

I would like to quote Rudolf Steiner's descriptions of the same spiritual event from a different point of view. He describes the flight of a swarm and how it gathers to form a swarm cluster. He connects it with the events of death and rebirth of the human soul.

> But a beehive is not a whole person. Honeybees cannot find their way into the spiritual world. We must bring them to another beehive to bring about their reincarnation. This is a direct image of a reincarnating human being. Those able to observe this have a tremendous respect for these swarming old bees with their queen, who actually behave the way they do because they want to enter the spiritual world. But they have become so physically materialized that they cannot. And then the bees snuggle together, becoming a single body. They want connection. They want to leave this world. Because they know: while usually they would be flying, now they gather on a tree trunk or the like, cuddle up together in order to disappear, because they want to enter the spiritual world. And then they will become a real colony again, when we help them, when we bring them back to a new hive.[8]

Animals of the Threshold

The previous depictions demonstrate to what extent honeybees are animals of transition. Their healing effect lies where the spiritual and earthly worlds touch and merge. Honey's harmonizing properties for humans is a small indication of the healing effects that honeybees have on the Earth and humankind on a grand scale.

The descriptions of the effects of bees on humans and the life conditions of the Earth are key to understanding the reasons for the current plight of bees. In our present human consciousness, the spiritual world is separated from the physical world as never before. The threshold between the worlds seems to have become mostly insurmountable in our human consciousness, which no longer experiences this threshold as an essential characteristic of being human. In fact, today, people have often lost the certainty that there even is another world, a spiritual world. Honeybees are beings whose mysterious tasks consist of keeping the transition between these worlds open, movable, and alive. But human beings are in the process of closing the gateways between worlds in many regions with all their might. Honeybees, as guardians of these gateways, are therefore falling ill. What people are creating in the world as the manifestations of their thoughts, feelings, and impulses of will, overwhelm the bees.

The task of honeybees is to reside and mediate between the spiritual and physical worlds. But this passage is getting more difficult, rigid, and hardened. The suffering of bees is caused by the fact that the different worlds, which belong together, are being separated by humans. This is what weakens the bees. The substances that poison bees are not the sole cause of their distress. Just like the poisonous substances themselves, the very thoughts that lead to creating these substances in the first place, followed by the use of these substances against one's better knowledge, all of this robs honeybees of their life forces and has a paralyzing effect on them.

Honeybees are completely exposed to human influences that shape Earth's organism with spirit and soul. On the one hand, bees can only do their healing work because they are so devoted to the human world. On the other hand, they have placed themselves in the custody of human beings and now largely depend on our protection and support. Honeybees are deeply connected to the impulses of human life.

Supporting and healing the bees therefore depends on the transformation of humans and their attitude towards the Earth and its beings. The healing of bees cannot be thought of and carried out without the self-healing of human beings.

The Varroa Mite

It is not my intention to discuss the treatment of bee diseases. However, in connection with the transformative work of honeybees, the being of the Varroa mite should be considered. Almost every bee colony in Central Europe is infested with this mite and therefore treated several times a year by beekeepers through a laborious process. Varroa mites develop along with bee larvae in the brood cells. When the bee hatches, the mites hatch with it and immediately begin to look for more brood cells with bee larvae in order to continue their reproduction. In one bee year, several generations of Varroa mites grow in a colony, with the result that their numbers are increasing progressively. Since these mites feed on hemolymph, the bees' blood, honeybees are weakened by a large infestation. If beekeepers do not take steps to reduce the Varroa mites, it is likely the colony will eventually die.

What is the relationship between the being of the Varroa mite and the honeybee? When I started looking at the mite, I was surprised that the honeybee has no dislike or hatred for it. I couldn't believe my "eyes," but I did not get the impression that the being of the bee was treating the Varroa as something foreign. There was no trace of antipathy in their relationship. This surprised me because I knew how badly Varroa-infested bees have to suffer. I was even more surprised by the answers I received from the honeybee being when I asked questions about the nature of Varroa mites.

A Varroa mite infestation only makes something visible that is already burdening honeybees. The mites show what from human beings is spiritually a burden for the bees—the shadows of the human soul, which people do not transform themselves, and which now strain the Earth and its beings, especially the bees.

Mites are not the cause of this burden. They are its expression. It therefore makes no sense to hate Varroa mites or even to associate feelings of antipathy with them. It is through them that evil is *revealed*.

The Varroa can be described as a shadow being that falls from the human soul onto the bees. It is the human soul's shadow made visible, which honeybees continually have to deal with and are no longer able to bear without human assistance. The presence of Varroa mites shows that the bees are burdened with more than they can endure.

The Varroa mite has roused beekeepers and made them aware that the bees are sick. With the emergence of Varroa, caring for honeybees has become a major challenge. New questions are being asked. People have started to think more about the relationship between humans and bees. Humanity's responsibility towards nature and the Earth has become an important issue.

One could say that Varroa mites have taken on pathogenic influences that are spiritually effective on Earth due to humans. They have actually prevented honeybees from getting sick more rapidly. The bees' suffering received a name through the mites. They enabled beekeepers to begin to deal with illnesses of honeybees. Since then, many bee colonies have been treated against Varroa. Honeybees suffer from the mites, but they are simultaneously protected from what the Varroa has taken on and what would have decimated them much more quickly. The Varroa mite is something like a buffer for the bees. However, the bees' current situation indicates that this buffer is slowly being used up.

HUMAN-BEE AND BEE-HUMAN

Soul of a Beehive

In his lectures about honeybees, Rudolf Steiner made a statement that can serve as an introduction to this chapter. He speaks here on the unity of beehive and beekeeper:

> If a beekeeper who is well liked by their colony falls ill or dies the whole of the colony really does descend into disorder. That is so. Well, one person whose thinking was right in line with current views said: "But the bees cannot see that well, they have no idea about the beekeeper. How is some sort of common bond supposed to arise? What's more, let us assume the beekeeper looks after the beehive in one year and next year there is a completely different bee colony in it; it is completely new apart from the queen; there are lots of young bees. How is the common bond supposed to arise in that situation?" I answered as follows: Anyone who knows anything about the human organism knows that it replaces all its substances within a specific period. Let us assume someone gets to know another person who goes to America and comes back ten years later. It will be a completely different person to the one whom they knew ten years ago. All the substances are different, the assembly is completely different. That is no different to what happens in a beehive in which the bees have been replaced but the common bond between the beehive and the beekeeper is

maintained. Such a common bond is based on the enormous wisdom which exists in the beehive. It is not just a heap of individual bees but the beehive really does have a concrete soul of its own."[9]

What is meant by this reference to the soul of a beehive? What is this soul? We will consider these questions in the following.

Winter Cluster and the Human Soul

In 2004, when I first began researching bees with the scope of suprasensory perception, I was repeatedly drawn to the mystery of the winter cluster. I heard the words of a Bienerich:

> The secret of honeybees is revealed in winter. You will discover their secret in winter, not in summer. In summer, they reveal the sensory side of their being, but in winter, they reveal their spiritual side.

Later he added words that preoccupied me for a long time: "In summer, human beings tend honeybees, but in winter, honeybees tend human beings."

In the time following, I tried to make sense of these words. They seemed to contain an essential secret pointing to the spiritual connections between humans and bees. The words describe that this connection is much more intimate than I had previously dared to think. I was very happy about this indication. But I also sensed that I had to work gently with it.

The winter state of a bee colony differs significantly from the state it assumes in summer. In summer, it is entirely devoted to collecting and transforming substances. The bees take care of the preservation and protection of the colony, raise their young, create honey, and ensure the pollination of many flowers. A bee colony is in constant motion and development. Even at night, it doesn't rest. In winter, however, this activity comes almost completely to a standstill. The colony forms a cluster that ensures its survival. Within this winter cluster, the bees are able to generate so much warmth that each individual insect can survive in spite of the cold. In the core of the cluster, the temperature can reach up to 30°C [86°F]. The bees are in a continuous, flowing movement, so that those on the outside can come to the inside and thus do not get too cold.

By meditatively viewing a winter cluster, we can get to know an important side of honeybees and touch upon a great mystery. One becomes aware that a colony is opened to the processes of the physical substances of summer due to the expansion it experiences. It works on specific substance transformations, supports the growth of plants, and provides people with high-quality food and medicine. A bee colony completely surrenders to the physical conditions of the Earth in the summer. In winter, the colony contracts, and physically occupies only a tiny space, which is many times smaller than the space it fills in summer. But this doesn't mean the colony is passive. It has to protect itself from the cold outside. It does so by converting forces present in the honey into movement and warmth. But one can also notice that a spiritual change is now taking place in the colony. It lets go of earthly conditions. By forming a self-contained entity, it expands spiritually.

Physically, a bee colony contracts in winter; spiritually, it expands. It expands in such a way that it is drawn to the spiritual core of the Earth. Each colony of bees becomes a small Earth in winter. It surrenders to the light that spiritually illuminates the Earth.

We can now direct our attention to this spiritual expansion of the bee colony in winter and ask about its significance. Here we make the surprising discovery that there is a connection between the condition of the winter cluster of a bee colony and the human soul. When the soul observes what is happening spiritually in a winter cluster, it encounters a living image. This image is deeply related to the soul itself.

What the soul recognizes in this spiritual image is the archetype of its own evolutionary path. By directing our spiritual gaze to the state of the winter cluster, we are led to the archetype of the development and maturation of the human soul. It is possible to read the soul's path of evolution in this primal image. The being of the honeybee has taken on the task of guarding the evolution of the human soul. This is deeply connected to the development of human freedom. In this sense, honeybees guard humankind.

The Honeybee Path of the Soul

In times past, people clearly understood that the beings surrounding them were their teachers. The animals and plants, and also the Earth as a whole, taught humanity, and their existence heralded mysteries of existential concern. They still do.

This included, and still includes, honeybees. The being of the Great Bee is not only pledged to bring forth bees from the spiritual world, but also carries an important task for the human soul and its development. The Great Bee guards the archetype of the spiritual path of human development. This archetype of the evolutionary path of the soul can therefore be called the path of the honeybee. The path of the honeybee is nothing else than the path of soul purification.

So, those who tend honeybees already follow a certain path of development. The way of the bees is given to them directly through the work on and with honeybees. However, this path is also available to people who do not keep bees. It is there for everyone, because it is a human path, and it becomes passable through the spiritual being of bees. Through the life of bees and their existence, the path opens up to human beings who strive for inner development.

Honeybees are tied to the development of the human soul. Therefore, we could also refer to this path as a particular path of the soul conveyed to the searching and listening human by the being of the bee. This path has a particular quality and direction because it comes to human beings from the bee being. Its principle is this: the human soul must completely renounce its own judgment when it resolves to understand a situation or a being. To truly understand an unfamiliar being, the human soul must abstain from its own peculiarities and only absorb what free, unhindered perception has to offer.

To the best of my ability, I will now describe the honeybee path. I understand that this is only one of the possible ways of describing this path. To do so, I am dividing it into seven stages. The individual phases of this path actually flow into one another imperceptibly. However, a division into seven stages assists in a deeper understanding. It makes it easier to describe this path.

This path is the path of the human being who wants to develop themselves. It is a path of development. People walking this path attain a deeper knowledge of the things and beings around them. Therefore, it is not only a path of development, but also a path of knowledge. The honeybee path represents the archetype of humankind's path to knowledge.

The first stage of the path can be called the stage of *devotion*. The human soul gives itself over. It opens itself, unreservedly, to what it wants to experience or understand. An inner gesture of surrender unlocks the soul. This way the soul is held back from passing a judgment before the being that the soul wants to understand is able to develop a life in the depths of this soul that is trying to perceive it. This is the phase of purposeful unintentionality. The soul undergoes a development through which it lets go of what would only cloud its own discernment. It holds back its own individuality, because only this way can the being it wants to understand assert itself. In this stage, the human being overcomes the tendency to avoid the unfamiliar, which usually is subordinated to what one already knows. The being one would like to understand loses its own dignity if it is only incorporated into the observer's already established system. However, its dignity is preserved when it can speak out of itself, and communicate with the language and means that are available to it. The stage of devotion, the first on the honeybee soul path, frees us from judgments that cloud our immediate vision and serves to open our understanding.

The second stage is the stage of *conception*. The soul receives into itself the being it wants to understand. It unites itself with the being in its awareness. However, this takes place without the soul being conscious of it. One cannot consciously experience this union. It happens when the soul is asleep. Only later does it wake up to what is unconsciously going on in this second stage of the honeybee path. People who strive for a conscious confrontation with the world may feel a bit put off by this phase. But even consciousness cannot deny the unconscious realms of the human being. It is justified to include them in the process of knowledge. The honeybee path is characterized by the fact that the human being is viewed in its wholeness. We have already gotten to know honeybees as beings that lead a life on the threshold between light and darkness. In the second phase of the honeybee path, the human soul is immersed in its own dark, unknown being. The dark, unconscious side of the soul comes into its own. In this way, human beings learn to reconcile themselves.

The third stage involves lifting what the soul has received into the *light of consciousness*. With their powers of consciousness, human beings now attempt to understand what they have received from the being to whom they directed their undivided attention. This stage is what could be called a slow, spiritual birth. The impression the soul has received is lifted from its depths into the light. What the soul received has been forgotten. Now, the conscious part of the soul begins to see what has sunk into it. It does this through the power of memory, which serves to bring into awareness what the soul has received. What it receives from the beings who step into the circle of its awareness is full of wisdom. The soul takes in the spirit of things and beings. But it absorbs them in a dreamlike way. In the third stage, we awaken the wisdom that has been bestowed upon our soul during the deep connection with the Earth and its beings.

In the fourth stage, we gain the ability to observe clearly what we have been able to bring up from our own soul with the power of memory. Here is the *encounter with a being*. What one's soul was able to call forth by moving through the previous stages, one can now observe as something external. This is necessary because otherwise one would not have the opportunity to name and describe what has passed through the soul and since departed. With the forces of the I one is able to look at what was brought out of the soul through the capacity of memory. We observe what is revealed of the being that had been taken in by the soul through the forces of devotion.

The fifth, and the two following stages, transform what has been brought about by the fourth stage. The goal is to more and more deeply understand the inner image that was created by the original impression. But the forces that the soul abstained from in the first stage, now come into play. It is about *understanding*, and from the fifth to the sixth level, the conscious human more deeply understands and permeates the spirit and order of wisdom expressed by the observed being.

The seventh stage is the conclusion of this path of knowledge. Human beings pass on what they were able to understand. They do not want to keep it to themselves because they realize that it would die. It is about *letting go* of what we have experienced. It is about forgetting it again, because only then are we ready to turn to another being.

Gifts of the Honeybees

I already mentioned that two puzzling words have been whispered to me by Bienerichs since the beginning of my research: "Human-bee and bee-human." My previous descriptions may have shed some light on the meaning of both these words. It is clear that bee colonies have a special relationship with human beings. If we really let what has been presented sink in, we can rightly claim that honeybees are soul siblings of human beings, who are sent to us from the spiritual world. We are accompanied by them on our earthly journey. They stand by our side in our tasks on Earth. The tasks honeybees take on for human beings have already been described in the previous chapters. I will summarize them once more.

The first gifts of the honeybees, belonging to the physical realm of life, are the tangible, substance related, transformative tasks they perform. These are undertakings that ultimately result in the creation of honey. In order to produce honey, there is intense contact between bees and the landscape in which they live. Honey's foundational constituents come from the flowers of the landscape. It can therefore be seen as an "imprint" of substance, an essence of the landscape. The connection that honeybees form with the landscape is very intimate. The spiritual being of the landscape lives in the plants and flowers. Nectar is a substance of the landscape that bees collect and use to form honey. Its healing forces for humans are naturally inherent in the process of its creation. Therefore, we can call honey the bees' first gift to their soul siblings, human beings.

The second, more spiritual gift that honeybees give to humans is the work of transforming the Earth and humankind. Through their activities, bees trigger a healing current that frees the Earth from its spiritual and psychic burdens. They also have a healing effect in the realm of sleeping people and those human souls leaving the earthly world.

The third gift relates most closely to the spiritual organization of the human being. It concerns the winter cluster of a bee colony. Through this gift, the honeybees provide human beings with strength for their spiritual development. The mysterious designation of the temple maidens as honeybees (Melissae) in ancient Greece is a reference to this third offering of the bees. The path of the honeybee queen through the wintry Earth, and her reappearance in spring, points to inner, spiritual forces of human beings. The queen's path through the night of the year has a mysterious connection with the path human beings travel when they search for a higher knowledge of Earth and existence.

Words of the Fairies

Fairies are high-ranking elemental beings who are connected to both the landscape and humans. They live in very specific places in the landscape, such as on the edges of forests, over streams, and in meadows or clearings. But one can also find them in parks or in significant places in the middle of cities. They are guardian beings who mediate between the spiritual life of the Earth and the existential being of humans. They have a wide-ranging, comprehensive awareness. People can come into contact with them and be instructed by them. Indeed, they even look for people who want to enter into a relationship with them.

In our context, it is interesting to hear what fairies have to say about honeybees. It's no surprise that there is a deep connection between them and bees; both fairies and bees are creatures of the landscape. The following words may appear strange to some readers, but I decided not to leave them out. On the one hand, they confirm the previous descriptions; and on the other hand, they add significant aspects.

> We fairies exist under the sign of the honeybees. This means we carry the mark of the bees on us. Every fairy is a bee in this sense. She is connected to bees and their mission and she acts in the spirit of the bees. That is why every fairy is a bee. Even those human beings who are connected with the mission of honeybees, and therefore work in their spirit, are bees. That is, they are called bees. From the standpoint of spiritual beings, naming and being are the same thing. One who is called a bee is a bee. In the spirit, there aren't separations like you are used to from your point of view, and that you find right and important to adhere to in your world. These separations do not apply to us spiritual beings. To us, naming and being are the same, even if it is completely different among you humans. And so, it can be that there are people who are bees because they live and work in the spirit of the bees. What kind of people are they, you may ask? They are people who are close to bees, even if they do not know it, and may not have anything to do with bees or insects. Nevertheless, they carry within them the way of the bees, the strength and the mission that honeybees pursue. People who are bees sometimes recognize themselves because they are working on behalf of bees. People who are close to bees share certain themes, life themes, with one another. But they also have certain convictions, specific guiding thoughts they want to realize in life. Some of these guiding thoughts are:

Let go of what you are overpowering, because it is constricted by your will.

Live and love your freedom and that of other beings above all else.

See if your gaze is able to draw out the deep essence of the beings that you meet.

Live in peace with your shadows. There is nothing more you can do for the peace of Earth. Because what grows from your making-peace-with-yourself does not grow according to your will, but according to the will of God.

Connect with the beings you meet by letting them know that you will remain without any judgment and that you will also not tolerate them forming judgements about you. Because there is no judgment that does not also contain lies. Instead of judging, it suffices to be more free, more attentive, more lively, and full of joy in your perception.

What others say cannot carry weight. The only thing that matters are your own experiences. But the richness of these must be recognized step by step.

Peace means being patient, awaiting, and supporting the diversity of humanity. The source of peace lies in the diversity, in the recognition of the value of individuality. The evolving I of the human being and nothing else is the source of peace on Earth.

Laws are not designed to promote freedom. They do the opposite because they prevent

> individuals from experiencing the moral power within themselves.
>
> People who come together because they love, appreciate, and embrace the diversity of humankind, work to increase the peace of the Earth and of creation.
>
> The one who is standing must not ignore the fear of falling, but must let it go. When fear falls, one can stand out of oneself.

These are convictions, with which people under the sign of honeybees can recognize themselves. They are then also very close to the being of fairies. But it is not enough for them just to have these convictions, they also want to realize them. Because they experience that this is not at all easy, they hold back from voicing their convictions. They tend to only talk about things they know from personal experience. They are therefore people who are reserved and humble. They watch themselves closely and suffer greatly if they fail to act as they imagined. We fairies love these people and stand by them wherever we can.

Worker Bees' Spiritual Gifts to the Beekeeper

It is very interesting to take a closer look at the relationship that exists between the beekeeper and worker bees in the colony. It is indeed the case that worker bees have an impression of the beekeeper who tends their colonies. It is also clear that neither the queen nor drones are capable of this direct perception of the beekeeper. Only the workers maintain this connection between the colony and the human being. However, they perceive the being of a person approaching them in a very dreamy way. They feel the aura of a human being. This means that between the aura of the colony and the aura of the human being there is a contact, which leaves a certain impression on the worker bees. That is why it is fair to say that colonies recognize their beekeepers. They feel them and they also feel when a stranger is approaching, though these experiences of worker bees do not reach the queen. To drones, beekeepers are meaningless. It's as if they don't exist for them.

One can observe that worker bees not only have an influence on their beekeeper when awake, but also when asleep. The soul of the sleeping person is usually in contact with the beings it met during the day. Except, in sleep, these encounters are of a purely spiritual nature. Since the soul is only faintly connected to the body during sleep, human consciousness does not understand what it experiences during the night. It can therefore be said that beekeepers dream with their bees, even if they have no memory of these dreams. Through these encounters, the spirit beings of the colony's worker bees are able to do something good for the beekeeper. In fact, the bees are doing work on the beekeeper's soul. They support the person at night who supported them during the day. At a certain point on its journey through the night, the beekeeper's soul is surrounded by the little spirits of worker bees who belong to the colony they care for. This also means, however, that non-beekeepers do not get to experience this particular nighttime work of the spiritual beings of worker bees. This gift is present for those humans tending bees. Sleep is a time of harvest for the beekeeper's soul.

If a beekeeper dies, the connection between them and their bees does not dissolve immediately. In this case, too, it is the beings of the worker bees who maintain the connection to the beekeeper beyond death. Especially in the three days after death, the workers' little spirit-souls are present with the one who died. It can be observed that they are active deep within the body of the deceased. Their effectiveness has to do with perceiving the individual internal organs of the body. One can see that the spirit bees clean and purify the organs and thereby restore them in such a way that the soul can more easily let go of the body. The little spirits work for their beekeeper in a very loving and careful manner. It seems as if they are giving back to the beekeeper the love that they felt for the bees. Only when the human soul has detached itself from the body completely do the little bee spirits disappear.

Beekeepers and Their Colonies

I have heard a beekeeper talk about how he goes into hibernation like his beloved animals—which his family regrets—but which he cannot seem to shake. Beekeepers often report how they wake up from sleep with a definite feeling that they absolutely must go visit their bees. And then they find some of the hives wantonly knocked over. Such examples, which occur often, show that an intensive exchange takes place between the beekeeper and the colonies being tended. One can rightly say that over time the bee colonies become part of a beekeeper's being. The bees settle into them in such a way that they get the feeling that they are working on themselves when they work on the beehives. Of course, relationships of various beekeepers are colored differently. But there is a foundational color, which has to do with the fact that certain spiritual beings participate in the relationship between the beekeeper and their colonies.

The first and most important being to be considered here is the Bienerich. It is the being through which the elemental world of a bee colony touches the human world and has a direct effect on the individual person, the beekeeper. To the human being, a Bienerich is a representative of a world, which as a secret, invisible world supports bee colonies in their work.

When a beekeeper opens a beehive, an encounter takes place between them and the Bienerich. As the beekeeper approaches the colony and interacts with it, the Bienerich is able to establish a deep connection with the beekeeper. The Bienerich can meet the beekeeper corporally. It can be observed that the Bienerich somewhat merges with the beekeeper while they are working on the colony, as if the two are melting together into one. A beekeeper can inwardly experience this encounter. They may notice that while working with the colonies, a change takes place in them, affecting their consciousness. Working on the colonies triggers a special state of consciousness. This is because a blending takes place between the Bienerich and the beekeeper. The human world and the elemental world touch and penetrate one another. This description may irritate or even cause fear in some people. One can get the feeling of losing autonomy as the foundation of one's life. But it is a spiritual reality that there are many encounters between humans and beings of the elemental world, which in most cases go unnoticed. The individual person is in constant contact with beings from the elemental world. As a rule, one only has a very faint perception of this. How the beings of the elemental realm interact with human beings and their individualities is a chapter of spiritual research unto its own. But I can say here that this relationship is closer than is generally believed and known. In the case of a Bienerich, this relationship is particularly close. Both the Bienerich and the beekeeper perform the same task, but each from their own point of view. They really belong together. It is therefore a welcome step for the Bienerich—and also the honeybees—if a beekeeper forms and lives into the relationship with the Bienerichs of their colonies more and more consciously. It is beneficial to the entire bee organism and the bees' tasks when the beekeeper occupies themselves with the connection that exists between the bee colony and the elemental world. It opens up a realm of experience that can mean significant progress in caring for bees. Humans are able to approach honeybees in a conscious way by getting to know their spiritual life conditions.

An Image of the Future Soul

Over and over, one can read about a comparison between bee colonies and human communities. Accordingly, the way bees work together should serve as a model for human society. After some thought, however, one will come to the conclusion that a bee colony cannot serve as a model for a human community, because the impulse of freedom does not exist in it. The individual bee is not free. It is tied into a natural process that fully determines its life and development. Therefore, it is not acceptable to apply the organization of bees to a human community. It is, however, much more appropriate to see an image of the human soul in the bee colony. In the honeybee colony, one comes face to face with a living image of the human soul.

This comparison can only be fully understood if one takes into account that this living image of a bee colony does not reflect the present soul, but the future state of the soul. Through honeybees, human beings can look ahead into the future of their soul. Future forces of the soul reveal themselves. These are already present in human souls, but only germinally. Through the honeybee colony, our future soul is reflected into the present time.

Something is present in every bee colony that indicates the future form of the human soul. Human beings are already permeated by it, but only in their sleeping will. They do not yet realize in themselves what the bee colony is already revealing. Due to the fact that most human beings have little or no knowledge of their own future being, they sleep through, so to speak, the moment they could become aware of this connection that exists between them and honeybees. Nevertheless, this lack of knowing allows them to turn toward the Earth in peace. They can devote themselves to their earthly tasks.

The future human soul, which reveals itself in a honeybee colony in the form of a living image, is deeply connected with the forces of the blood. These forces connect human beings with the past, with their families, parents, tradition, language, and their folk. Everyone is standing in this stream and is therefore a part of it. This stream of the past serves our individuality in the sense that we can evolve in it. It offers our individuality the foundational requirements for our own development, but nothing more. The step that leads to an individuality beginning to become active out of itself is always simultaneously a rejection and transformation of blood forces. The I is not found in the past. There, one can only determine to what degree one's development has progressed so far. But one is not shown the direction in which to proceed. So, the I must transform the blood forces in order to begin becoming aware of its own powers. I have a strong impression that the process by which a bee colony produces honey is precisely an image of the transformation the I must undertake in the blood to assure itself of its own powers.

For human beings to become free, they must liberate themselves from the binding forces of the blood. But it would be a mistake to think that one can do without these forces that make one an earthly human. It would be wrong to simply deny these blood forces. This is not even possible, because they are very strong forces that do not allow us to ignore them. Rather, human beings have the task of transforming them, by redeeming what is bound and trapped in the blood. Forces of a pure, redeemed blood are revealed through a honeybee colony. Materially, this purified blood can be seen in honey. Human beings are called to acquire forces of transformation. The image of the Holy Grail represents this pure, redeemed blood. For the searching human being who has decided to walk the path of personal transformation, this sacred vessel is present. As soon as one has made this decision, one is nourished by the Grail. One's soul receives comfort and healing from it. The various transformative tasks of honeybees that I have described are ultimately indications of the healing effect that the Grail has on the human soul. In this sense, honeybees are animals of the Holy Grail.

In honeybees—more precisely, in a bee colony—a force, an essence, shows itself, which points to the future of humanity. Within this essence, a state is revealed that is yet to come for the human soul. Strictly speaking, a bee colony is held together by forces that human beings will one day have at their disposal. They will then have integrated the wisdom of honeybees into themselves as a living force, as a capacity. The archetype of the bee colony has a deep connection with humankind. But it is human beings as they will be in the future, who are related to this archetype. This connection explains the kinship between honeybees and humans. When human beings are devoted to honeybees, they are devoted to forces that will be freely available to humankind in the future. Beekeeping serves the future capacities of humankind.

Through the honeybees, a future, purified human soul has received a temporary abode on Earth, according to its present living conditions. Through bees it has a dwelling place on Earth. Honeybees remind human beings not to forget of what our own evolution consists.

Honeybee Verses

The annual developmental cycle of a bee colony is made up of certain fixed stages. Many valuable observations and insights have been gathered about the sequence and nature of these stages. The deeper meaning of these developmental stages of bees becomes apparent by observing sense phenomena as manifestations of a spiritual language. In the swarming of a colony, in the winter cluster, and in the queen's marriage flight, essential aspects are expressed. Each stage of development is an aspect of the spiritual archetype of a bee colony. This archetype is a very wise spiritual being. By studying these individual stages of bee colony development, one can find access to this exalted being.

The following verses point out a path in this direction. If one grasps the contents of these verses meditatively, one can acquire an understanding of the larger context of which these individual developmental stages are a part. This way, one can know and understand more deeply the being of honeybees, their life and rhythms.

The first seven of the twelve verses relate to the bees' annual cycle, from the swarm to the winter cluster. After that, the view widens to other important aspects of the honeybee being. The last five verses are about the connection between honeybees and their beekeeper, their connection to the landscape, their remedies, their Christmas mystery, and the death of a bee colony.

Bee Being

Swarm

Germinating seed of life
beginning of a new existence
blossom of a life to come
expand yourself
breathe
to the edges of the circle of life
create yourself anew
a wheel of life begins
the germinating seed is alive

Swarm Cluster

Out of depths
of the Earth
resounds
I am
I will be
I will have become

Moving In

Peace
envelops us
like cosmic sheaths
to which our life belongs
from which we receive our task
to become a home for erring human souls
just as we have become at home on Earth
after times of journeying
to inhabit hearts
opened to us
by human beings

Day and Night

    Harmony
is a gift from bees
    to the Earth
lifting the Fall
    of humankind
forming a center
    for the dead
healing the unbalanced

Marriage Flight of the Queen

Streaming light
sheaths of life
depths of being
she rushes through
suffering a Sun death
to be born anew
from the being
whose message she carries
life-promising
being-unifying
Earth-heralding
bride

Overwintering

The fruit of a whole year
lies among us
we are swimming in memories
of the living light of summer
we are nourished by
what the Earth has given
its streaming life
keeps alive
protects us
unites us
and strengthens

Winter Cluster

Falling
into a spiritual ocean
in infinite layers of Earth
receiving words
of a new birth
for her hive
now is the mother queen's life

Beekeeper and Bees

Sleep
slowly transforms to awakening
 in the realms of living light
  seeing the being
   from which they hail
    from which they emanate
     to which they return
     in constant change
    of life
   of growth
  of flourishing
 and ripening
So they stand before me
as the guardians
of my own soul
so I stand before them
as their caretaker

Bees and Landscape

Infinite protection
in the light
to fly through it
we were born
to warm it
we were sent
nourishing
unifying
arranging
the life of Earth
through our whole being
as animals of Earth
heeding the light
bound to transformation

Healing Remedies of Bees

PROPOLIS
gives protection, peace, and rest
HONEY
Inclines
to receive
life laws of higher order
WAX
envelops life
in living, warming light
POLLEN
nourishes
what is starving
VENOM
enlivens
what is too close to death

Honeybee Light

In darkest night
a quiet light awakes
calmly flowing
lovingly blessing
wants to be lifted
seen and accepted

It is the life light of the innermost soul
speaking out of the dark Earth
warmly enveloping
filling with strength
Receive it
pass it on
for the Earth is in need

Death and Transformation of Bee Colonies

What passes
is lifted
into a new state
begins a new existence
lies down
dissolves
turns to dust
its real life
mysteriously
living on
in different form and shape
It is resurrected
always renewing

Death is not an end
life expands to new shores
when it passes
they are twins
of each other
  realizer
  refiner
  overcomer

Quiet is hope
solidifying over time
for fruits appear
later
but unmistakable to one
reading the signs
imprinted by life
on the eternal

Honeybees as Messengers of the Spiritual World

Messages from Honeybees

This chapter is a collection of messages I was able to receive from the spiritual beings of honeybees in 2005. They contain indications that may contribute to a deeper understanding of beekeeping. They point to spiritual connections existing between bees, human beings, and the Earth. These texts have accompanied me over the past few years and have become an important wellspring for my work in the honeybee world. I now wish to make them known to others. The last two messages are of later dates, from November 2012 and June 2013.

Beekeeper Consciousness

Take seriously what you are doing and its effects on the Earth. Do not underestimate the value of the motives underlying your actions. Searching for these motives, and the resulting conscious handling of them, is an essential field of activity connecting you to the hidden being of honeybees. Streams, soul streams, flow back and forth between the deep concerns of humans and the high being of the honeybee. Underestimating these causes damage to bees. Earlier generations knew about or sensed these streams. Beekeeping was and is above all a sacred task, yes, a work with the most holy. Beekeepers were unwaveringly aware of this in earlier times. They knew their work, their encounter with bees, with all of nature, to be a service like that of a priest at the altar. They met beings of nature like a priest meets the Godhead. And that is how it shall be again.

For the Beekeeper

Learn to observe bees with your soul. Empathize with them, so that the boundaries disappear between what you perceive to be outside of you and what you know and experience as qualities within your soul. Observe in a way that stimulates you to make discoveries within yourself. But do not do this in a manner where you think too much; do it with your feelings, your empathy. Move your soul into what you see, into what you experience, without letting yourself be distracted by any knowledge. Surrender to these impressions and unite with them by breathing them with your soul. But do not think you have to surrender completely. Have courage to experiment and make self-observations of your soul when you are with honeybees. Sense yourself as a student in the school of honeybees, as an explorer of your own soul through what bee life stimulates in you. This is how you learn to listen. This is very precious and truly indispensable if you wish to work for the bees and not against them. You can only succeed by developing something rather rare among people, that is, real listening. A listening with your soul and not your ears. Then human beings are listening as beings aware of their own soul and spirit, taking this awareness with them into the encounter, into the work with honeybees. So, do not set aside the truth of your great, expansive human beingness, but educate it and let it gather experience while working with bees. Expand with the bees and do not restrict yourself and thereby the bees. Grow with the bees, learn inner growth, learn new things, truly new things, and do not force yourself and the bees to live in the opposite, the narrowness of customs and rehearsed knowledge. The being of the honeybees challenges humankind to follow a path of inner growth. This path always includes taking risks and overcoming oneself to face one's own shadows and patterns. This is necessary. It is unmistakable that bees are the very beings with the capacity to constantly create anew. Working with bees in a way that supports their well-being requires that human beings spiritually find this capacity within themselves and bring it into practice for the bees. Otherwise, it is inevitable that they will be harmed.

School of Love

Human beings are not yet what they will be. But they are on their way. It is important that humans see themselves as beings who are on the way. And with them the whole Earth is on its way. As they walk this path, they receive assistance that they may or may not notice. One of the beings approaching them is the honeybee. Bees bring the essence of love to people on the path. One can understand the deep mission of honeybees when one understands them as beings in which love is manifested as a living image. When a human being listens to the bees like a studious apprentice, they can foresee: the love of humankind will one day be what the unity of honeybees is now. Sensing this, human beings are in the school of love. They can further see: without the being of bees, love could not exist on Earth. Bees are the guardians of love. Until human beings have matured to bear it in their own souls, bees will be the keepers of love. Until then, the honeybee being is assisting humankind in the preservation of love.

Gifts of the Earth

The admirable social life of bees is bestowed upon them by their innate presence in the realms of the inner Earth. Honeybees carry up from these realms what has already been established in the Earth by Christ. That which is resting in the Earth as a spiritual mystery, affecting nature, yet hidden from human beings, can be seen in bees as visible form. The most sacred soul sheath, belonging to bees because they reach so deeply into the realms of the inner Earth, is an element that human beings will acquire only at a future time. It is a form of the holiest, purest astrality, connecting humankind to future incarnations of the Earth. Honeybees are the carriers of this future element through their activity and through their interweaving with both the entire Earth organism as well as the sacred cycle of the year. As a gift from the Earth, honeybees are an expression of the golden kingdom of love at the center of the Earth. Beeswax is the expression of an event that in the Earth corresponds to the transition of the minerals from a stage of solidity to a stage of liquifying, refining, softening. The inner essence of mineral substances emerges. Mineral substance reveals its inner being, which is by no means as hard, coarse, and solidified as the visible shell suggests. This process, actually taking place inside the Earth, as a process of dematerialization, is directly expressed in beeswax. Wax formation begins where mineral turns into what is living, flowing, dissolving. Formation of honey indicates a process one step further than the formation of wax. Honey is the expression of an earthly event that consists of pushing living matter so far that its foundational spiritual substance can come into appearance. The essence of matter appears in great purity. The inner realm of the mineral is revealed. Honey is an expression of this. Going beyond what is revealed in wax and honey, a deep mystery of the Earth is present in the social life of honeybees. Something most holy is expressed in the social life of bees. But it rightly appears as an intensification of what already resounds in wax and honey, because, like them, it is connected to the mysteries of the depths of the Earth.

Plight of the Earth

The Earth is in need. She needs your help. She aches under the weight that lies upon her and the burdens that were and still are being imposed on her by humans. Human beings must create spaces, soul spaces, inner spaces of warmth and light. This is an art, an art of living, the creation of inner spaces of liberation. Wherever these are not continually created anew, the Earth is breaking. The Earth is expressing her distress, which is actually a distress of the human beings who live on her. The bees work and work to purify the Earth because in her there are zones which are stifled, zones of consumption and of desolation. She must be freed from these. Human beings can only assist in this work if they do so from an experience of inner peace, of inner freedom. To assist in illumining the Earth, they must discover their own inner light. Action is necessary.

Earth Gold, Nerve Processes, Bee Gold

In the depths of the Earth lies a realm of spiritual gold, which has a special effect on human beings because it envelops their nerves when they are thinking. To what extent this Earth gold can exert its effect on them depends on their thoughts, their thinking. It can have a soul-ennobling effect when people find the courage to enliven their thinking, that is, to allow the same process of evolution that the Earth and humanity are undergoing to flow into the structure of their thought process. The gold of Earth can thereby be effective in human nerve processes. The day-to-day building up and regeneration of nerve substances are only possible because people become partakers of the gold of Earth while they are sleeping. Elemental beings rise from the Earth into the human brain during sleep, and spiritual Earth gold is worked into the nerves, so that these can regenerate and be available again to human consciousness. Elemental beings clothe nerve substances in the gold of Earth.

As beings of the inner Earth, honeybees are obliged to ennoble the Earth organism through the active forces of inner Earth gold. They are particularly connected with Earth's Christmas. This is why they rest in winter, because their being enters into the Earth to be united there with the Christmas light. Honeybees' participation in Earth's Christmas is a prerequisite of their ability to accomplish their healing work for the Earth. Earth gold can be felt right down into the substances the bees produce. The healing process occurring through honeybees is related to the process contributing to the regeneration of human nerves. This is how a bee colony is related to the human head.

The Light of Earth

The Earth is nourished from its center by the realm of the spiritual Sun. The Earth lives from this center. Honeybees bring forth the spiritual light of the inner Earth's Sun. It lives in them because they live in it. Their home is the spiritual light of the Sun in the Earth. Human beings also live from this spiritual Earth light because they are beings of Earth. It would be a big developmental step for human beings if they were able to take account of everything they owe to this spiritual light of the inner Earth. They would not be able to live if it were not for the effect of this light. It is their spiritual source of life. That honeybees bring up this inner light is one of their deepest mysteries. Because they do this, they reside in Christ, deep within Him, for His light rests within the Earth. Honeybees are therefore messengers of Christ. When a bee dies, it flies into the Christ. He welcomes every soul of a dying bee. Before it is born again, it is allowed to live in the garment of the Christ Being, to crawl here and there, and be nourished by His Being. The bees are His children. For He has made them so. Honeybees perform a transformative task on Earth. This they do in connection with higher, superior spiritual beings. Their activity consists of a constant permeation of the Earth's body. They do this in service to the Being through whom they are what they are. Their life is in service to Christ.

Bee Death

Each bee has a rich, immeasurable life that humans can hardly imagine. If they could feel this life, they would no longer disregard bees and no longer feel that they are the masters who rule over bees. They would find new relationships through which honeybees could connect with them. Honeybees once gave themselves into the hands of human beings, into their care, not as subordinate creatures, but as sister beings. One cannot rule over one's siblings. No bee has life experiences only for itself. They have them for their colony, for the higher spiritual beings connected to honeybees, and for the Earth. This, then, is the foundation of the bee's own development. In every bee, there is an elemental being. Through the bee, it becomes a little soul. This elemental being once entered the circle of bees and will leave this circle again when the time comes. Such are many cycles that connect nature beings to one another. Before it enlivens a bee, this elemental being is connected to the world of plants. For a long time, it works its way through the various realms of the elemental world that connect plants to their surroundings, from the mineral realm, through the water realm, to the realm of air and light. This elemental being, which is destined to become a bee soul, ascends within the plant to the blossom. There it can transform and become the soul of a bee. It does this under the guidance of higher spirits connected to the honeybee being. In the same way, a bee soul leaves the circle of bees as soon as its time has come and is given new tasks under the guidance of other, higher beings. The experiences it has as a bee are transformed into skills that serve to carry out new tasks.

Honeybees in September

In September, honeybees feel their way into the spiritual realm of the Earth. They are then at the threshold between the outer light of the Earth visible to human beings and the spiritual light of the Earth. To bees, the Earth is slowly becoming transparent. Their attraction to the outer world of light, air, and warmth decreases. Instead, what one could call the realm of the spiritual support of the Earth begins to appear. The effects of the Earth itself increase. Honeybees belong to this inner kingdom of Earth. Much more than they are beings of the above-ground Earth, which they busily buzz through in summer, they are spirits of the inner Earth. Honeybees are children of the Earth and its hidden realms. The mystery of bees and many other secrets of life cannot be understood by human beings if they do not give the innumerable beings of the inner Earth a worthy place in their thinking. Yes, the Earth will fall if human beings do not allow these essentials into their thinking. They deprive themselves of the foundation of life if they cannot recognize that the Earth is an organism and is therefore inhabited, enlivened, formed, and transformed by a multitude of spiritual beings. Honeybees are beings that move back and forth between the realms of the outer and inner Earth. This is why they are important to humankind.

A soul of a dying bee slips into the Earth, leaving behind its physical shell. Its path leads it into the various realms of the inner Earth. The bee soul takes along a spiritual form of the physical shell until the damage that this spiritual body has suffered through its existence above ground has been repaired. The restored spirit body then remains in the upper layers of the Earth. It serves as an orientation for the bee souls pushing toward birth, in order that these soon-to-be-born souls can find their way back to their aboveground home. The ascending souls clothe themselves in these spiritual bodily forms, which enables them to develop into a new bee animal in the honeycomb. The recently deceased bee's little soul, having left its spiritual body behind, enters a region in which all descending souls gather. They gather around a spirit that can be called the soul mother of bees. She takes in the souls, and they experience comfort and gentleness. All the heaviness that they brought with them from their existence above ground is stripped away. These souls experience a belonging to the being of the Great Bee, regardless of which colony or queen or which place on Earth they belonged to. They discover their interrelatedness through an expanded spirituality connecting to the spirit of Earth. The spirit being of the inner Earth appears to them as an open secret. They become witnesses to the inner Earth light.

Honey Mystery

Honeybees distribute gold over the Earth, Earth gold. It flows into the substances which bees pass on to human beings as gifts. Like all substances, they have a spiritual side. This means that they have an effect on people through their very existence. This is why honey has an effect even if it is not consumed. Something flows into the honey that is only present in small traces in other substances. It is the essence of peace, balance, and equilibrium that lives in honey. No other substance is made the way honey is made. A great, superior weaving of essence flows into honey. The healing properties of honey reveal themselves in the process of its creation. It also expresses the highest tasks of honeybees. In honey, the mystery of the being of bees appears as substance.

Honeybee Christmas

Honeybees are beings of the innermost Earth. It is the home and origin of their spiritual being. This is why beekeepers have always sensed a mystery between their bees and the events of Christmas. Because Christmas is a festival of the innermost Earth. Honeybees are a living image of the spiritual organs of the center of the Earth that reaches all the way up to the physical surface of the Earth. In the center of the Earth, one finds the Earth's fountain of youth, its sacred source of life and its life-giving heart. As beings of the Earth's center, honeybees bear a work of redemption, through which sacred, healing, spiritual substance is lifted up for humankind and the beings of Earth.

Words of the Bees

The being of the honeybee
stands by humankind unconditionally
as humans must yet learn to do themselves
In this regard, they are not yet
where the bees already are

Despite fragmentation
of one's own soul
to be able to find oneself as a human being
is the mysterious gift of the bees
to humankind

In honey
the sacred being of Earth is living
has become substance
honey contains no fear
All fears and doubts are foreign

In it they are overcome
Bees are their conquerors
In their flying and doing and being
lies the power of transformation
that appears
in honey
in the world of substances

The drones love the Earth
the worker bees the darkness
which they transform
the queen the human soul
which she precedes in the spiritual world

Honeybees fly through the hearts
through the blood of human beings
Their work for humankind
is a work of the heart
is a transformation of the blood
the darkness of human blood
the forces of destruction
of denial
the ignorance

the egoism
are what
they fly against
fighting
for these powers are in blood
spiritually present
The resistance of bees can be found
where human beings are exposed to evil
Their battle does not escape the bees
They can do nothing other
than assist
in the transformation of opposition
faced by the human soul

The spiritual queen speaks
to human beings
of their highest most sacred forces
of transformation
Spiritual living spaces of development
are created by worker bees
for the human soul
The spiritual spark of life
nourishing all growth
is carried by drones from the spirit world
continuously
into the physical world

Receiving of life
yet to be lived
individual humans actualize
the bee being in themselves
Transforming the old into the ripeness of the new
they are living
what the bees are doing
through the power of their individualities

Aware of the signs
appearing in the outer
they realize the human-bee
Their soul becoming aware of its bee being
through awareness of the signs
continually sent
from the spiritual world
This is their nectar

All human beings are whole
otherwise they would not exist
They just do not yet have the resources
to recognize this
Honeybees work spiritually toward this realization

Do not search
but let yourself be found
by the beings
who wish to speak to you
who work
on the completion of Earth
through time
through lifetimes
through experiences
that are present
at every moment
at every place
and work
in the paths
of human beings
across the Earth

Worker Bee, Queen, Drone

We are waiting for human beings who listen to their own soul like a beehive. Every soul has a scent, a sound, a potential, and is in a certain state of expansion or contraction, just like a bee colony.

The workers are the bearers of the Christ force, which flows out and in, brings in and carries out, divides and reunites, and thereby participates in the bee colony's transforming of the shadows through the light. Because the soul is subdivided into many, many aspects of being, because it is multi-layered and varies from moment to moment, it can transform the experiences that it has and that rest in it, as past and future impulses of life.

But the Now is given by the drones, their modesty, their dark, warm, calm Earth tones open the soul to the world of the Now. Their presence in the soul alone, as the power of the calm, eternal Earth, enables the soul to access the Now. Without them, the soul could not bear the Now.

And the queen! She is the heavenly in the human soul, the heavenly light of human beings, or the star of their existence, their birth, which, although so bright and light, sinks down the deepest. But this is how the life stream, the spark of the I can nestle in the human soul. Nothing moves as deeply into the dark night of Earth as the being of the honeybee queen. When she is born spiritually, she almost dies in the process. When she receives the drones, she almost dies in the process. In winter, she moves into the holy Midnight Sun of Earth and is called to a new life by the Lord of the Earth Himself.

Listening to the beehive, the soul meets itself as a cosmic child in earthly garment.

About the Illustrations
by Karsten Massei and Franziska van der Geest

Honeybee Cosmos (p. 12)
These are the spiritual spaces encompassing each bee colony; the spiritual habitat a bee colony opens up between the light of the outer Sun and the light of the inner Earth. (K. M.)

Spiritual Beings of Bees (p. 18)
Various spiritual beings appear in this picture, who surround and guard each bee colony. (K. M.)

Honeybee Queen Born from the Earth (p. 28)
Here is shown the event of the gradual birth of the queen out of the Earth in the time after Christmas. The path is shown in the ascending structure on the left. Beings of the Earth who accompany this process are visible. The light of the Easter Earth can be seen at the top left. (K. M.)

A Swarm (p. 74)
A swarm comes flying in from the top left. The elongated, hovering, blue structure represents the swarm cluster. The coexistence of the bees in the swarm cluster is due to the memory of the time they spent with their queen as a colony. The inclination to build a new honeycomb structure is within the archetype of the bee colony. When the colony has found a suitable place, they start to build immediately.

On the right edge, one can see a material sheath (beehive, tree cavity) that a bee colony needs in order to build its new home. During swarming, the honeybees are carried and guided from all sides as if on hands of very tall beings, who appear everywhere in the picture. Spiritual beings continuously support the swarm along its way. (F. v. G.)

Bee Being Among Transformative Processes of the Animal Realm (p. 96)

The care of the bees by human beings during all stages of life can be seen in this picture. From the bottom right, this path of steps climbs upwards to the left in a large lemniscate movement through the center of the queen's heart. The queen's blue form indicates her deep connection to the etheric stream of life present around her. On the left are the elemental helpers who let honey flow into the comb. The developmental stages of the brood are shown to the right of the queen in her wing area. Behind that, the animal guardian stands in a purple robe, responsible for all the animals. Under the heart of the queen, an animal head with horns appears, representing the ruminants. This indicates the importance of the queen's digestive system, which is responsible for egg formation.

The dark figure at her feet is a human being searching for truth, preceded by a small brown bee being. The bee is gifting them remedies, which again give them a protective cloak, shown in the large, brown figures on the left. Thus, the bee being has completely united with the human being, and through the remedies the human being can recognize its own deepest being. The light force that had been tied up in the human being and made it appear black and white is transformed by the remedies into a body that has been warmed-through, so that the soul takes on the color of Earth. The honeycomb region shown at the bottom left, into which the honey flows, is also the birthplace of bees. The dying of the bees is shown in the lower right corner. The moment a bee dies, the decision to reincarnate is already made and a new cycle begins. The bee bodies are of great importance to the

Earth. The Earth drinks in and lives from the adventures and experiences that bees have gained in their short intense lives. Each body creates a new piece of Earth that never existed before. They are therefore a model to us of the great circle of life. (F. v. G.)

Bee Being (p. 128)

In the picture, one can see the auric periphery surrounding bees as a suprasensory sheath. In the middle, one can see the elongated queen bee, surrounded by a bright glow, but wearing a dark dress that ultimately ends in the venomous stinger. In the bright glow, the influence of various other animal beings and their powers can be experienced. At the top edge of the picture, three spirit beings are depicted, which embody the three bee beings (queen, drone, worker bee). However, they also have a connection to human beings, in that soul forces are revealed in them through which they remain part of the cosmic body of the Earth. A Bienerich appears to their left. Below it, a sphere becomes visible showing which connections it has to the spiritual organism of the inner Earth. On the right edge of the picture, one can see the forces necessary for drones to be created. Next to it (orange shape) are the forces through which the workers come into being. The queen being, from which drones and workers descend, appears in the center. At the bottom right, a beekeeper is resting and their soul is listening to the bees. In the lower left half of the picture, there are animal and plant beings that are protected in the cave of Mother Earth. Bee venom is ingested by a holy woman. (F. v. G.)

Honeybees as Messengers of the Spiritual World (p. 142)

This picture reveals the path of the soul, which it follows after death, in the bee being. (K. M.)

Human being
in us appears to you
the goal of your evolution
the spirit spark
the fullness
whose carrier
you are

Notes

1. Karsten Massei, *School of the Elemental Beings* (USA: SteinerBooks, 2017)

2. Rudolf Steiner, *Bees* (CW 351) (USA: Anthroposophic Press, 1998), lecture 8, December 22, 1923, Dornach, p.157.

3. Michael Weiler, *The Secrets of Bees: An Insider's Guide to the Life of Honeybees*, translated by David Heaf (Great Britain: Floris Books, 2006), p. 29.

4. Ibid., pp. 26–27.

5. Ibid., pp. 35–36.

6. Steiner, *Bees*, p.156.

7. Weiler, *Secrets of Bees*, pp.48–49.

8. Steiner, *Bees*, p.157.

9. Rudolf Steiner, *Understanding Healing: Meditative Reflections on Deepening Medicine through Spiritual Science* (CW 316) (Great Britain: Rudolf Steiner Press, 2013), lecture 1, January 2, 1924, Dornach, p. 12.